# The Happy Brain

## The Science of Where Happiness Comes From, and Why

DEAN BURNETT

First published by Guardian Faber in 2018

Guardian Faber is an imprint of Faber & Faber Ltd,
Bloomsbury House, 74–77 Great Russell Street,
London WC1B 3DA

Guardian is a registered trade mark of
Guardian News & Media Ltd,
Kings Place, 90 York Way, London N1 9GU

Typeset by Faber and Faber Limited
Printed and bound by CPI Group (UK) Ltd, Croydon, CR0 4YY

A CIP record for this book
is available from the British Library

ISBN 978–1–78335–129–9

2 4 6 8 10 9 7 5 3 1

The
Dean Burnett is comedian and a currently a lect edic F...

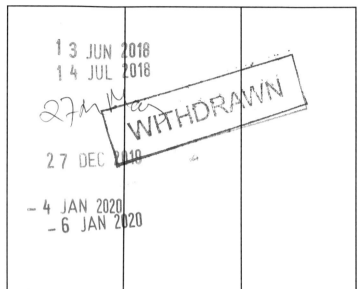

a neuroscientist, ...
...thor. He is 35 and liv...
...urer/tutor at the Cardiff U...
's ...cation. His first book, *The* ...
...al bestseller published in ov...
...apping' blog is the most read on t...
..., with over 15 million views since...

*To everyone who bought my first book.*
*This is all your fault.*

# Contents

# Introduction

As a wise philosopher once said, 'Happiness, happiness, the greatest gift that I possess.' Aristotle, I think. Or possibly Nietzsche? Sounds like something he'd say. No matter, the point is valid; happiness is important.

But what makes anyone happy? Why are different people made happy by different things, and at different times? What's the *point* of happiness? *Is* there one? The reason I was interested was because I was meant to be writing a second book, but I had no idea what it should be about. Everyone I asked gave different suggestions, but eventually always said, 'Just write about what makes you happy.' As a very literal, scientifically minded type, I tried to look this up: what *does* make us happy? But all I found was an avalanche of management fads and techniques, cod philosophy, self-help manuals, life coaches and gurus, all of varying degrees of dubiousness, and all insisting that they definitely knew the secret to happiness, no matter who you are. I wouldn't mind so much, but barely any of these 'secrets' matched up, suggesting that a lot of them might be *nonsense*.

Case in point, here are some real headlines from the UK's notorious *Daily Mail* newspaper: 'Forget cash – how sex and sleep are the key to happiness'; 'Key to happiness? Start with £50k a year salary'; 'Why the secret to happiness is having 37 things to wear'; 'Is treating yourself like a baby the key to happiness?'; 'Key to happiness for over-55s? Buying a new pet and going for a day trip with lunch at a pub every month'; 'The key to happiness? Handing out cakes on the street'; and so on. Make of that what you will.

Even more annoying for a doctor of neuroscience, science writer and apparent go-to guy for mainstream commentary on brain-based news like me, is that a lot of these so-called secrets invoke my discipline, or constantly refer to some valid-sounding-but-unspecific aspect of the brain's functioning, like 'dopamine' or 'oxytocin' or 'emotion centres', in support of their claims. If you're an experienced neurobod, you can easily spot when someone is just borrowing the terminology of your field to sound credible, rather than actually having any useful understanding of it.

And I thought, you know what? If you're going to exploit my field, at least put some *effort* into it. Sure, the brain isn't perfect, I'm often the first person to point that out, but it's still one of the most fantastically and terrifyingly complex things to study. To truly explain how the brain deals with happiness would take more than a vague two-line summary or a smattering of impressive-sounding terminology, it would take a whole book . . .

And that's when it dawned on me. *I* could write that book! The one about how the brain really handles happiness at the fundamental levels. And that's the book you're holding now. Because if there's one thing I do, it's go to ridiculous extremes to settle minor grievances, even if the party that caused them remains blissfully unaware of my existence.

So, this is a book about happiness and where it comes from in the brain. What causes it, and why? What makes our brains like certain things so much, but not others? Is there some sure-fire way of inducing happiness in any human brain like so many seem to claim, suggesting that happiness is like tapping a password into an online bank account? Can eternal happiness actually exist – and would it be desirable, anyway? Wouldn't experiencing the same thing day-in day-out for years on end be more likely to drive you to the edge of madness than provide everlasting satisfaction? And more.

One thing that is abundantly clear from the sheer variety of supposed 'secrets' to happiness is that it has an undeniably strong subjective element. We all have different ideas of what makes, or will make, us happy, be it wealth, fame, love, sex, power, laughter, and so on. And yet we can only ever truly know what works for *us*. So, I wanted to include insights from a wide range of people from different walks of life, to see what makes them happy (or not). As a result, I ended up talking with stars of stage and screen, millionaires, leading scientists, journalists, ghost-hunters and one person who . . . well, let's just say that in no other research I've done did I ever hear the term 'sex dungeon' used so freely and so often.

I should warn you though, that this is *not* meant to be a self-help book, or some model for how to live a happier and fuller life, or anything like that. I'm just fascinated by the brain and all that it does, and one of the things it does is allow us to experience happiness. It was my intention to explain, to the best of my abilities, how it does this. I hope you're happy with that. Although if you're not, I'll understand why.

And once you've read the book, so will you.

# 1

# Happiness in the Brain

Would you like to be stuffed into a tube? Head first?

Don't answer yet, because there's more.

Would you like to be stuffed head first into a tube, a cold and confining one, where you're not allowed to move? For hours at a time? A tube that makes incredibly loud noises, an ongoing din of clicks and screeches like an enraged metal dolphin?

Pretty much everyone would say no if asked this question, before hurriedly seeking out the nearest authority figure. However, imagine not only agreeing to this, but actually *volunteering* for it. Repeatedly! What sort of person would do that?

Well, me. Yes, I've done this many times. And I would do it again if asked. I don't have a weird and incredibly specific fetish, but I am a neuroscientist, a keen student of the brain and a science enthusiast, so in the past I've volunteered for various neuroscience and psychology experiments. And since the dawn of the current millennium, many of these experiments involved having my brain probed by fMRI.*

MRI stands for Magnetic Resonance Imaging, a complex hi-tech procedure which uses powerful magnetic fields, radio waves and several other types of tech-wizardry to produce

---

* Admittedly, I did make it sound far worse than it is for comic effect. You can make any everyday experience seem terrifying by creative use of language, e.g. 'Would you like to be stripped naked and jammed in a hi-tech coffin that will bombard you with harmful radiation?' sounds like a terrible experience, but sunbeds are very popular nonetheless.

very detailed images of the inside of a live human body, revealing things like broken bones, soft tissue tumours, liver lesions and alien parasites (probably).

But more attentive readers will have noticed that I referred to *f*MRI. The 'f' is important. It stands for 'functional', so it's functional magnetic resonance imaging. This means that the same approach used to look at the structure of the body can be adapted to observe the *activity of the working brain*, allowing us to witness the interactions occurring between the countless neurons that make up our brains. It may not sound that impressive, but this activity is essentially the basis of our mind and consciousness, in much the same way that individual cells make up our body (cells combine in complex ways to form tissues, which combine in complex ways to form organs, which combine to form one functioning entity that is you). Scientifically speaking, this is a fairly big deal.

But . . . why am I telling you this? We're supposed to be looking at where happiness comes from, what's with the detailed description of advanced neuroimaging techniques? Well, while it would be dishonest of me to deny that talking about complex neuroimaging methods does indeed make me happy, there is a much simpler reason.

You want to know where happiness comes from? Well, what is happiness? It's a feeling, or an emotion, or a mood, or a mental state, or something like that. However you define it, it would be extremely hard to deny that it's something that is produced, at the most fundamental level, by our brains. So there we go, happiness comes from the brain. That's everything wrapped up in a page, right?

Wrong. While it is technically *correct* to say that happiness comes from the brain, it is also essentially a meaningless statement. Because, using that logic, *everything* comes from the brain. Everything we perceive, remember, think and imagine. Every facet of human life involves the brain to some degree. Despite massing just a few pounds, the human brain does a ridiculous amount of work and has hundreds of different parts doing thousands of different things on a second-by-second basis, providing us with the rich detailed existence we take for granted. So *of course* happiness comes from the brain. But that's like being asked where Southampton is and replying 'the solar system'; correct, but utterly unhelpful.

We need to know precisely *where* in the brain happiness comes from. Which part produces it, which region underpins it, which area recognises the occurrence of happiness-inducing events? For this, you have to look inside a happy brain, and see what's happening. It's no simple task, and to have any hope of doing it, you need sophisticated neuroimaging techniques, like fMRI.

See, told you it was relevant.

Unfortunately, there are several obstacles to this particular experiment.

Firstly, a decent MRI scanner weighs several tons, costs millions and produces a magnetic field powerful enough to pull an office chair across the room at lethal speeds. And even if I could get access to this super-machinery, I wouldn't know what to do with it. I've been *in* one many times, but that doesn't mean I know how to operate one, any more than taking a long-haul flight means I'm a pilot.

My own neuroscientific research was into behavioural studies of memory formation.[1] While this may sound impressively complicated and detailed, it mostly involved constructing elaborate (but cheap) mazes for lab animals to solve, and watching how they did it. All very interesting, but it means I wasn't trusted to operate anything more dangerous than a box cutter, and even then most people would leave the room, just in case. I was never allowed near anything as elaborate as an MRI scanner.

My luck was in, however. I live a very short distance from CUBRIC, the Cardiff University Brain Research Imaging Centre, where I volunteered for all those studies. It was being built as I completed my PhD at the Cardiff Psychology School, and was opened just after I left. This timing seemed a bit mean-spirited if I'm honest, like the whole institution had said, 'Is he gone? Good, now we can break out the good stuff.'

CUBRIC is an excellent place to go for the latest cutting-edge investigations into the workings of the human brain. And, doubly lucky for me, I have friends who work there. One of these friends is Professor Chris Chambers, prominent expert and researcher in brain imaging techniques. He was happy to meet with me, to discuss how I planned to go about locating happiness in the brain.

However, this would be a business meeting, not a social one. If I wanted to convince a professor to let me use his incredibly valuable equipment to pursue my personal investigation into how the brain processes happiness, I needed to make sure I'd done my homework. So, what does science already know, or suspect, about how happiness works in the brain?

## Chemical happiness

If you want to know which bit of the brain is responsible for happiness, consider what counts as a 'bit' of the brain.

Although it's often thought of as a single (surprisingly ugly) object, it can be broken down into a vast number of individual components.* The brain has two hemispheres (left and right), made up of four distinct lobes (frontal, parietal, occipital, temporal), each of which is composed of numerous different regions and nuclei. These are made up of brain cells called neurons and numerous other vital support cells called glia, which keep things functioning. Each cell is essentially a complicated arrangement of chemicals. So you could say that, like most organs and living objects, the brain is a big lump of chemicals. Chemicals arranged in breathtakingly complex forms, but chemicals nonetheless.

In fairness, we could break it down even further. Chemicals are made of atoms, which are in turn made of electrons, protons and neutrons, which are in turn made of gluons, and so on. You end up getting into complex particle physics as you delve deeper into the fundamental makeup of matter itself. However, there are certain chemicals the brain uses for purposes beyond basic physical structure, meaning they have a more 'dynamic' role to play than just being the building blocks of cells. These chemicals are neurotransmitters, and they play key roles in the functioning of the brain. If you're looking for the most simple, fundamental elements of the brain that still have profound impacts on how we think and feel, these chemical neurotransmitters would be them.

The brain is essentially a huge and incredibly complicated mass of neurons, and everything the brain does is dependent on, and the result of, patterns of activity generated in these

---

* Just to be clear, at no point should you literally attempt to physically break a brain down into its components. This will mean immediate death for your subject and life imprisonment for you.

neurons. A single electrochemical signal, a pulse known as an 'action potential', travels along a neuron and, when it reaches the end, is transferred to the next one in line, until it reaches where it's meant to go. Think of it like an amp* travelling along a circuit from a power station to your bedside lamp. It's quite an impressive distance for something so insubstantial to travel, but it's so common we barely even consider it.

The pattern and rate of these signals, these action potentials, can vary enormously, and the chains of neurons relaying them can be incredibly long and branch off almost endlessly, allowing for billions of patterns, trillions of possible calculations, supported by connections between almost every dedicated region of the human brain. That's what makes the brain as powerful as it is.

Stepping back slightly, the point at which the signal is transferred from one neuron to the next is incredibly important. This occurs at synapses, the point where two neurons meet. However, and here's where it gets slightly strange, there's no significant physical contact between the two neurons; the synapse itself is the gap between them, not a solid object. So how does a signal travel from one neuron to the other if they don't touch?

Neurotransmitters is how. The signal arrives at the terminus of the first neuron in the chain, and this causes the neuron to squirt neurotransmitters into the synapse. They then interact with dedicated receptors in the second neuron, and this causes the signal to be induced again in that neuron, and it's then relayed along to the next one in line. And on it goes.

Think of it like an important message, sent by the scouts

---

* As in 'ampere', the basic unit of electric current, not 'amplifier', the big boxy devices for making musical instruments louder. That would just be confusing.

of a medieval army to the commanders back at headquarters. The message is on a piece of paper, being carried on foot by a soldier. He reaches a river, but needs to get the message to the camp on the other side. So, he ties it to an arrow and fires it across, where another soldier can pick it up and carry it further along the journey back to headquarters. Neurotransmitters are like that arrow.

The brain uses a wide variety of neurotransmitters, and the specific neurotransmitter used has a palpable effect on the activity and behaviour of the next neuron. That's assuming the next neuron has the relevant receptors embedded in its membrane; neurotransmitters only work if they can find a compatible receptor to interact with, a bit like a key only working for a specific lock, or series of locks. To go back to the soldier metaphor, the message is encrypted so only those from the same army will be able to read it.

There's also a wide variety of orders the message could contain: attack, retreat, rally forces, defend the left flanks, and so forth. Neurotransmitters are similarly flexible. Some transmitters increase signal strength, some reduce it, some stop it, some cause different responses altogether. These are cells we're talking about, not inert electrical cables; they're diverse in how they react.

Because of the diversity offered by this setup, the brain often uses specific neurotransmitters in certain areas to fulfil certain roles and functions. So, with this in mind, is it possible that there is a neurotransmitter, a chemical, responsible for producing happiness? Surprising as it may seem, this isn't that far-fetched. There are even several candidates for such a thing.

Dopamine is an obvious one. Dopamine is a neurotransmitter that fulfils a wide variety of functions in the brain, but one of the most familiar and established is its role in reward and pleasure.[2] Dopamine is the neurotransmitter underpinning all activity in the mesolimbic reward pathway in the brain, sometimes called the dopaminergic reward pathway in acknowledgement of this. Whenever the brain recognises that you've done something it approves of (drunk water while thirsty, escaped a perilous situation, been sexually intimate with a partner, etc.), it typically rewards this behaviour by causing you to experience brief but often intense pleasure triggered by the release of dopamine. And pleasure makes you happy, right? The dopaminergic reward pathway is the brain region responsible for this process.

There's also evidence to suggest that dopamine release is affected by how *surprising* a reward or experience is. The more unexpected something is, the more we enjoy it, and this seems due to how much dopamine the brain deploys.[3] *Expected* rewards correspond with an initial dopamine surge, which then tails off. But *unexpected* rewards correspond with an increased level of dopamine release for a longer period after the reward is experienced.[4]

To put this in a real-world context, if you see that money has arrived in your account on payday, that's an anticipated reward. Conversely, finding £20 in an old pair of trousers, that's unexpected. The latter is much less money, but it's *more* rewarding, because it wasn't expected. And this, as far as we can see, causes a greater dopamine release.[5]

Similarly, *absence* of an expected reward (e.g. your pay isn't in your bank account on payday) seems to cause a substantial *drop* in dopamine. Such things are unpleasant and stressful. So, obviously, dopamine is integral to our ability to enjoy things.

But as mentioned previously, supporting pleasure and reward is just one of dopamine's many and varied roles and functions across the brain. Perhaps other chemicals have more specific roles in inducing pleasure?

Of course, endorphin neurotransmitters are the 'big daddy' of pleasure-causing chemicals. Whether they are released from gorging on chocolate or due to the rush of sex, endorphins provide that oh-so-wonderful intense giddy warm sensation that permeates your very being.[6]

The potency of endorphins should not be underestimated. Powerful opiate drugs like heroin and morphine work because they trigger the endorphin receptors in our brains and bodies.[7] They're obviously pleasurable (hence the alarming number of people who use them), but these drugs are also clearly debilitating. Someone in the grip of an intense opiate 'high' isn't much good for anything other than staring into space and occasionally drooling. And some estimates suggest that heroin is *only 20 per cent as potent* as natural endorphins! We have substances five times as powerful as the most intoxicating narcotic just hanging around in our brains – it's a wonder we get anything done at all.

While it's bad news for pleasure seekers, it's good news for the functioning of the human race to hear that the brain uses endorphins very carefully. Most typically, the brain releases endorphins in response to serious pain and stress. A good example of both is childbirth.

Mothers use many terms to describe childbirth – 'miraculous', 'incredible', 'amazing', and so on – but 'enjoyable' is rarely among them. And yet despite the extreme physical demands it places on a woman's body, they get through it, and often do it again. This is because human women have evolved many different adaptations to facilitate childbirth, and one of these is the build-up and release of endorphins as it progresses.

The brain deploys endorphins to dampen the pain and stop it from reaching heart-stopping levels (which can happen[8]). This could also contribute to the almost deliriously happy state women experience the moment the baby is born (although that's possibly just relief). Thanks to endorphins, childbirth, no matter how gruelling it is, *could be worse*.

That's one extreme example. There are other ways to expose yourself to enough pain and stress to trigger an endorphin release (like by being a man and telling mothers that childbirth could be worse). Putting your body through other sorts of physical extremes, for example. People who do marathons report the 'runner's high', an incredibly pleasurable rush that occurs when your body is physically taxed enough for the brain to break out the big guns and drown out all the aches and pains.

It could therefore be argued that the function of endorphins isn't to induce pleasure, but to prevent pain. Maybe labelling endorphins as 'pleasure inducing' is like describing a fire engine as 'a machine that makes things wet'; yes, it does that, but no, that's not what it's *for*.

Some argue that this agony-reduction function only applies to *detectable* levels of endorphins, where their action is noticeable to the person.[9] There's evidence to suggest that at a lower concentration endorphins play a more basic role, helping regulate behaviour and task management. The endorphin system, via complex interactions with the neurological systems that regulate stress and motivation,[10] helps us know when something is 'done'. An important task needs doing and you get stressed; you complete the task and the brain releases a subtle dose of endorphins so we feel 'it's done, let's move on'. Not exactly producing pleasure, but helpful and reducing stress, thus contributing to wellbeing and happiness.[11] This is further evidence of the preventative function of endorphins in maintaining happiness.

One problem with both the dopamine and endorphin explanations is they assume 'happiness' is the same as 'pleasure'. While it's certainly possible (normal, even) to be happy while experiencing pleasure, to be truly happy surely requires a lot more than that. Life is more than just a series of euphoric moments. Happiness is also about contentment, satisfaction, love, relationships, family, motivation, well-being, and many other words found in Facebook memes. Could there be a chemical that supports this more 'pro-found' stuff? Maybe.

One contender would be oxytocin. Oxytocin has an unusual reputation, often being described as the 'love' hormone, or the 'cuddle' hormone. Despite what much of the modern media would suggest, humans are a very friendly species, and usually actively *need* social bonds with others in order to be happy. The closer and more intense these bonds, the more important they are. The bonds between lovers, relatives, very close friends, tend to make people happy over the long term. And oxytocin is apparently integral for these.

Going back to the process of childbirth again, oxytocin's most established role is as a chemical released in high doses during labour and breastfeeding.[12] It is key for this most fundamental of meetings between individuals – it causes the immediate and intense bonding between mother and baby, is present in breast milk, and induces lactation.[13] However, oxytocin has since been implicated in a much wider variety of situations: sexual arousal and responses, stress, social interaction, fidelity, and no doubt much more.

This has a number of weird consequences. For instance, oxytocin is important for forming and enhancing social bonds but it is also released during sexual intercourse. This may be why the oft-referenced 'friends with benefits' arrangement (where two friends opt to be physically intimate without any stifling relationship/commitment) is so notoriously difficult to maintain. Thanks to oxytocin, sexual interaction can fundamentally alter your perception of your partner, changing purely physical attraction into genuine affection and longing. Oxytocin is what's 'making the love' during lovemaking.

And while oxytocin affects women more than men, it does still have potent effects on men; for instance, one study showed that, when dosed with oxytocin, men in relationships will keep more of a distance between themselves and attractive women in a social context than single men do.[14] The conclusion drawn here is that increased oxytocin makes men more committed to their partner, making them more aware of how their actions might impact on them, meaning they'd be warier of interacting with unfamiliar attractive women, especially when others are there to see it. Basically, it can be argued that oxytocin strengthens existing romantic bonds. But it doesn't *create* them per se, hence single men don't show similar behaviour.

There is far more that could be said, but the point is that oxytocin is vital for the human brain to experience love, intimacy, trust, friendship and social bonding. All but the most cynical souls would agree that such things are crucial for lasting happiness. So, therefore, is oxytocin responsible for happiness?

Not quite. As with most things, oxytocin has a down side.

For instance, increasing your social bonds with an individual or a group can increase your hostility to anyone outside that bond. One study found that men dosed with oxytocin were much quicker to ascribe negative traits to anyone not from their culture or ethnic background.[15] Or, to put it another way, oxytocin makes you racist. If racism is integral to happiness, then I'm not sure humans deserve it.

It doesn't have to be so extreme though; you've probably witnessed someone (or even been that someone) experiencing bitter jealousy and resentment, even hatred, when the object of their affection is seen to interact in an overly friendly way with someone else. The fact that 'crimes of passion' exist shows just how potent and destructive this reaction can be. There are many ways to describe someone gripped by jealous rage or paranoid suspicion; 'happy' isn't one of them. Oxytocin may be crucially important for social bonding, but not all social bonding leads to happiness. It can, in fact, lead to the opposite.

Perhaps this whole approach is too far removed? Pleasure and intimacy could be said to *lead* to happiness, so any chemical that gives rise to these things is only indirectly 'causing' happiness. Is there any chemical that makes us happy directly?

Serotonin may do this. It's a neurotransmitter used in a wide variety of neurological processes, so has a diverse range of roles, such as enabling sleep, controlling digestion, and, most relevantly, regulating mood.[16]

Serotonin appears to be vital for allowing us to achieve a good mood, aka 'be happy'. The most prescribed antidepressants available today work by increasing the levels of serotonin available in the brain. Current wisdom argues that depression arises due to reduced levels of serotonin, and this is something that should be fixed.

Prozac and similar medications are classed as SSRIs, or selective serotonin reuptake inhibitors. After being released into the synapses to relay signals, serotonin isn't broken down or destroyed, instead it is re-absorbed by the neurons. SSRIs basically stop this re-absorption from occurring. The result is that rather than a quick burst of activity in the next neuron produced by a brief appearance of serotonin in the synapse, this activity is prolonged because the serotonin hangs around, intact, constantly triggering the relevant receptors. You know when your toaster gets old and keeps popping the bread out before it's done, so you have to leave it in for longer to get it how you like it? It's a bit like that. And this treats depression. Therefore, serotonin is obviously a chemical that causes happiness, right?

Not right. The fact is, nobody really knows (yet) what it is that the increase of serotonin is actually *doing* in the brain. If it's simply the case that there's insufficient serotonin to produce a state of happiness, then that should be an easy fix. However, given the speed at which our metabolisms and brains work, SSRIs increase serotonin levels pretty much immediately. And yet, most SSRIs take *weeks* of regular doses to be effective.[17] So, clearly it isn't just the serotonin itself which is responsible for a happy mood, it must be having an indirect effect on something else.

Perhaps the real problem is with the approach; you can attribute powerful neurological properties to simple molecules all you like, it doesn't mean that's how things work. If you look around, you can find many an article or column explaining how to hack into your 'happy hormones' or similar, claiming that a few simple diet and exercise techniques can raise the levels of the relevant chemicals in your brain, resulting in lasting contentment and enjoyment of life. Sadly, this is severe oversimplification of incredibly complex processes.

Essentially, it seems that trying to pin happiness on a specific chemical is the wrong approach. They're involved, but not a *cause*. A £50 note is valuable, and is made of paper. But it's not valuable *because* it's made of paper. And so it may be that the chemicals described here are to happiness what paper is to money; they allow it to exist, but their role is mostly incidental.

## Go to your happy place

So, if it's not caused by specific chemicals, where in the brain might happiness come from? Is there a specific *area* in the brain that processes happiness? A region that takes the information from other parts of the brain about what we're experiencing, assesses it, and recognises that it should make us happy, and so causes us to experience this much-sought-after emotional state? If the chemicals are the fuel, could this specific area be the engine?

It's certainly possible, but we need to be careful before jumping to any conclusions, and here's why.

As I write this (mid-2017), it's a good time to be a neuroscientist. The science of the brain and how it works has very much entered the mainstream, with major well-funded brain projects being announced in the US and Europe,[18] countless books and articles exploring the workings of the brain, regular news stories about the latest brain-based breakthrough or discovery, and so on. Exciting and lucrative times for neuroscience indeed.

But there are downsides to this mainstream popularity. For instance, if you want to report something in a newspaper, it has to be understandable to the readers, the vast majority of whom won't be trained scientists. As such, it needs to be simplified and stripped of jargon. It also has to be succinct, and this is truer than ever before in today's extremely competitive, attention-seeking, soundbite-craving media. If you've ever read any scientific publications, you'll know most scientists do not write in this way, so translating impenetrable technical reports of meticulously planned experiments into easily understood copy means a lot of changes have to be made.

If you're lucky, these changes will be made by a trained science journalist, or an experienced science communicator; someone who understands the requirements of mainstream platforms but grasps the information well enough to know what's important and what can be edited out in the name of clarity. Unfortunately, very often it isn't someone like this. It might be a less-experienced or underqualified journalist at a newspaper, or even an intern.* Or it could be the press department at the university or institute behind the research, who want to get publicity for their work and efforts.

Whoever it is, they'll often make changes or cuts that twist or even misinterpret the actual story. When you consider other factors that would distort the actual information (exaggeration to gain attention, emphasising of one particular issue by a newspaper with a specific ideological axe to grind, and so on), it's no surprise that a lot of science stories you see in the

---

* Science news is still often considered rather 'niche' by many mainstream platforms, so is often dealt with by people lower down in the hierarchy. I once had to do an interview to help out someone covering a science story for a major UK newspaper. The poor bewildered chap admitted he'd been working on the entertainment section until only the week before.

news are quite far removed from the actual experiments that produced them.

With something like neuroscience, a subject that gets a lot of coverage and interest but where the underlying science is quite messy, still relatively new and poorly understood, these distortions can lead to widespread, oversimplified ideas about how the brain works.[19]

One of these that keeps popping up is the idea that everything the brain does has a specific 'area', or 'region', or 'centre'. We see stories about the areas of the brain responsible for voting preferences, or religion, or enthusiasm for Apple products, or lucid dreaming, or overuse of Facebook (I've seen all of these in print). The idea that the brain is a modular mass, composed of clearly defined separate components each with a dedicated function (like an Ikea cupboard but slightly less confusing) is ever more pervasive. But the truth is more complicated.

The theory that certain bits of the brain are responsible for specific functions is centuries old, and has quite a disturbing history in parts. Consider the practice of phrenology, the theory that the shape of the skull can be used to study an individual's personality traits.[20] The logic is quite straightforward. Phrenology argued that the brain is a collection of dedicated thinking regions working together. Every thought or action or characteristic has a specific location in the brain, and, like muscles, the more a region is used or the more powerful it is, the bigger it is. So, for example, if you're smarter, you'll have a bigger region that processes intelligence.

However, when we're young our skulls are still malleable, gradually hardening as we age. According to phrenologists, this means that the shape of our brains influences the shape of our skulls, with larger or smaller brain areas resulting in bumps or dips in the skull. And these, they believed, can be assessed to determine the type of brain, and therefore the abilities and personality, of an individual. Someone with a more sloped forehead would be of low intelligence, someone with less pronounced bumps at the back of the skull would lack artistic ability, stuff like that. Simple.

The only real problem with this approach is that it was devised around the early nineteenth century at a time when having robust, thorough evidence to support your claims was more of a 'nice idea' than standard practice. Phrenology doesn't work at all. The skull may indeed be 'softer' when we're very young, but it's still several plates of relatively dense, sturdy bone, evolved to protect the brain from external forces. And that's not even taking into consideration the fluid and membranes that are wrapped around the brain as well.

The idea that minor variations in the size of brain regions, composed of spongy grey matter, could cause measurable distortions in our unyielding skulls that correspond with personality traits, reliably and in every individual, is ridiculous. Luckily, even at the time phrenology was a fairly 'alternative' science, and it was gradually discredited and fell out of fashion. Good thing too; it was regularly used in very unpleasant ways, like 'proving' white people were superior to other races, or that women were intellectually inferior (they are typically smaller and have correspondingly smaller skulls). This, coupled with the lack of mainstream scientific acceptance, gave phrenology a very unsavoury reputation.

One less obvious but still negative consequence of phrenology is that it set some contemporary neuroscientists against the theory of brain modularity, the notion that the brain has specific parts to do specific things. Many scientists argued that the brain is more 'homogenous', undifferentiated throughout its structure, so every part of the brain is involved in every function. Certain bits doing certain things? That sounds like *phrenology*, so any theory that hinted at this risked being met with cynicism.[21]

This is unfortunate because we now know that the brain *does* have many specific regions for performing certain functions. It's just that these regions are for more fundamental things than personality traits, and they certainly aren't detectable via lumps in the skull.

For example, there's the hippocampus, in the temporal lobe,* which is widely agreed to be integral for encoding and laying down memories; the fusiform gyrus, believed to be responsible for face recognition; Broca's area, a complex and diverse region of the frontal lobe responsible for speech; the motor cortex, at the rear of the frontal lobe, which oversees conscious control of movement. The list goes on.[22]

Remembering, seeing, talking, moving: all fundamental processes. But, to bring it back to the central point, could there be a brain region responsible for something more abstract, like happiness? Or, like phrenology in the past and mainstream media distortion in the present, is this an oversimplification of brain structure, taken to illogical extremes?

There is some evidence to suggest that assigning a brain region for happiness isn't so ridiculous. A number of regions seem to deal with specific emotions. The amygdala, for

---

* To clarify, the brain is composed of a left and right hemisphere, as previously stated. One hemisphere is usually 'dominant', hence people being left- or right-handed, but both are pretty much the same, structurally. So when I mention any particular area, like 'the hippocampus', the brain actually has two; one left, one right. The parallel regions often work together, or back each other up. There's a lot of redundancy in the brain. But it's easier to refer to them in the singular in this context.

example, is a small area next to the hippocampus crucial for giving memories an 'emotional context'.[23] Essentially, if you've got a memory of something that scared you, it was the amygdala that added the fear to that memory. Lab animals without an amygdala don't seem able to remember that they should be afraid of certain things.

Another example would be the insular cortex, situated deep in the brain between the frontal, parietal and temporal lobes. One of the functions attributed to the insular cortex is processing the sensation of disgust. It shows activation in response to noxious smells, sights of mutilation or anything similarly viscerally unpleasant, and even is believed to be more active when you notice an expression of disgust on someone else's face, or even when a disgusting thing is just *imagined*.

So there are two bits of the brain that process what many would consider a feeling, or emotion, much like happiness. Is there an area that is responsible for happiness itself?

One candidate was mentioned earlier; the mesolimbic reward pathway. This is found in the midbrain (a deeper, more 'established' area of the brain, down among the brainstem) and is responsible for providing the rewarding sensation experienced when we do something pleasurable. When it comes to happiness, as opposed to pleasure, some studies show that the ventral striatum needs to be active for lasting happiness. Others show that the left prefrontal cortex is elevated during feelings of happiness.[24] Another study argues that it's the right precuneus.[25] Basically, top scientists have looked for which bit of the brain produces happiness, and come up with a different answer each time.

This isn't as weird as it might sound. The brain is an incredibly complicated place, and the techniques for studying it in such detail are still, in scientific terms, relatively new. The idea of using rigorous analytical approaches and advanced tech to study intangible emotional states is even newer still. This means that the 'best' or 'correct' way to isolate happiness is still being sorted out, so you would expect some confusion and inconsistency at this stage. It's not the scientists' fault though (well, not usually), because there are many issues that confuse matters.

The most obvious is the method employed by the researchers to try and make their subjects 'happy'. Some use questions and instructions prompting happy memories, some use pleasing images, others use messages and tasks to induce a happy mood, and so on. Exactly how happy these can be said to make people is anyone's guess, and it no doubt varies considerably from person to person. And on top of that, the experiment typically depends on the subjects reporting how happy they are. This adds another layer of confusion.

It's a problem encountered by many psychology experiments that hope to analyse what humans do in certain contexts, under laboratory conditions. The fact is, being in a laboratory undergoing an experiment is not a normal situation for most people and so they tend to be a bit confused, and possibly intimidated, by it. This means they are more likely to do as they're told by the nearest authority figure. This is invariably the researcher, and subjects end up unconsciously telling them what they *think* they want to hear, rather than what the researcher *genuinely* wants to hear (in this case, an as-accurate-as-possible description of their internal state). There's always the risk the subjects are trying to 'help' by exaggerating or modifying the description of what they're actually feeling (e.g. 'This experiment is about happiness, so if

I don't say I'm happy I could ruin the whole thing'). Despite best intentions, this achieves the opposite of helping.

Taking all this together, it is clear that looking for happiness in a person's brain is fraught with challenges. We could get round these, though, if we could somehow get a subject who was totally familiar with the laboratory environment, who wasn't intimidated by researchers or their weird contraptions, who knew enough to be completely accurate in their reporting of their internal state, who could come up with their own experiment and even analyse their own data . . .

That settled it. I wouldn't just ask Professor Chambers if I could use his MRI machines; I'd ask if *I could be the one scanned*. It made perfect sense; I'd know if I was happy or not, and the situation would be far less likely to influence me, making any readings genuinely valid and informative. So all I needed to do was slide inside a scanner, switch it on, get myself into a happy state, and then look at the data. Job done.

Of course, once I'd come up with this idea, I was immediately hit by worries that it was ridiculous, or just plain weird. Luckily for me, even a cursory look at the body of research into happiness shows that things often get very strange indeed.

## Happiness is hard to find

In early 2016, I saw a talk by Professor Morten L. Kringelbach, head of the Hedonia: Transnational Research Group. Imagine if Benedict Cumberbatch played an accomplished Danish scientist. That's basically Professor Kringelbach. Except shorter.

Professor Kringelbach's Hedonia research group is a collaborative effort between Oxford University in the UK and

Aarhus University in Denmark.[26] They study the various ways people experience pleasure, particularly how it relates to health and disease. On this day, Professor Kringelbach was talking about something strange they had discovered.

The researchers were looking into what it is about some music that we enjoy so much that it compels us to dance. Many people enjoy dancing, and many people enjoy seeing it. Dancing makes a lot of people happy. But not everyone. Some people just don't like the idea of doing it, not where anyone can see them anyway. But even for these people, there are certain songs or tunes that compel them to move, even if the dancing is just a rhythmical tapping of the leg, or nodding of the head, or an unintentional shimmy when they think nobody is looking. If it's something people actively dislike, why would they still do that?

As Professor Kringelbach explained, there's a specific spectrum of musical properties that the brain prefers. The group's experiments show that there needs to be a medium level of syncopation (or unpredictability) in music to elicit a pleasure response and associated body movement in a person. What this means in plain English is: music needs to be funky, but not *too* funky, for people to like it enough to make them want to dance.[27]

Your own experience will probably back this up. Simple, monotonous beats aren't really entertaining (try dancing to a metronome and see where that gets you). They have low levels of syncopation and certainly don't make you want to dance. In contrast, chaotic and unpredictable music, like free jazz, has very high levels of syncopation and rarely, if ever, entices people to dance. Of course, some people will disagree with this, but then no matter how unpleasant/bizarre/

unfathomable something is, you'll find a human somewhere who likes it. They're good like that, people are.

The middle ground (funk music like James Brown is the most referenced by the researchers, and is also what Professor Kringelbach danced to for our considerable enjoyment) hits the sweet spot between predictable and chaotic, for which the brain has a strong preference. Most modern pop falls somewhere within this range. This is likely why you can hate a modern pop song with a passion, openly declaring you detest every single thing about it, and still find you're tapping your foot along with it when you hear it played in a shop.

The point is, for some reason, tunes that have a specific balance between predictability and chaos induce pleasure in our brains, making us happy to the point where we're compelled to physically respond. Clearly the underlying processes by which our brains determine what makes us happy are not exactly straightforward. It's not a simple yes/no matter of something making us happy, or not; often it's a *specific amount* of something that makes us happy, and any more or less has the opposite effect. Think of it like salt; too little salt in your food, and it doesn't taste nice. Too much salt in your food, and it doesn't taste nice. The right amount of salt in your food, it tastes good, and the poor waiter can finally move on to the next table.

Here's another weird finding; it might not even be our brain that determines our happiness, but our gut. While a number of clichés and sayings acknowledge links between our brains and digestive system ('the way to a man's heart is through his stomach' or 'I can't think on an empty stomach', and so forth) it might still be surprising to know there's a lot of scientific evidence to suggest the workings of our gut could have a direct and profound impact on our mental state.

It's important to remember that our stomach and intestines aren't just simple wobbly tubes that the useful bits of food pass through; they're incredibly sophisticated in their own right. As well as possessing a dedicated and intricate nervous system of its own (the enteric system, which can in some cases operate independently, hence it's often labelled 'the second brain'), our gut is also home to tens of *trillions* of bacteria, of thousands of different strains and types. All of these have potential roles to play in our digestion process, by determining the substances that enter our bloodstream and travel to every part of our body, potentially influencing the activity of every organ and tissue. Overall, it's clear these bacteria have direct impacts on our internal state.

Remember, the brain, despite its sophistication and baffling complexity, is still an organ. It's not just affected by the things we sense from the world outside our heads, it's also beholden to what's going on inside the body. Hormones, blood supply, oxygen levels, the countless other facets of human physiology: these all impact on the workings of the brain. Given that the gut (and the bacteria it's home to) has a crucial role regarding what goes into the body, it's perhaps to be expected that it would have significant, albeit indirect, influence over how the brain functions.* Scientists recognise this fact, and have coined the term 'the gut–brain axis'.[28]

One consequence of this convoluted relationship is that the gut has been strongly linked to occurrences of depression.[29] Some studies suggest that possessing certain strains and types of gut bacteria is a prerequisite of experiencing

---

* Before you think this is a one-way relationship, rest assured that the brain often dominates and overrules the digestive system, in many surprising and often harmful ways. Much of this is covered in my first book, *The Idiot Brain*.

stress and depression, and similar mood disorders.[30] Much of this evidence is limited to animal models at present, so it's tricky to say if there's such a 'profound' link between gut and mood in humans, but it's not that far-fetched.

Ninety per cent of the body's serotonin, the neurotransmitter seemingly crucial for being in a good mood, is found in the gut. We've also looked at how certain neurotransmitters determine our mood and perception of pleasure. These neurotransmitters are created in the neurons, and for this the neurons need a reliable supply of the substances and molecules used to manufacture the neurotransmitters. These building blocks are typically derived from the food we eat, and the bacteria present in our gut are integral to this. So, if we lack, or have too much of, the type of bacteria required to extract the metabolites (the component elements of larger complex chemicals derived from metabolic processes) for production of neurotransmitters, then the amount of this neurotransmitter available for our brain would be altered. This would surely affect our mood – or so you would think.

While this 'gut bacteria affects our mood' claim is reasonable to a certain extent, it overlooks the fact that it's an *incredibly* complicated arrangement and system, and this brief description doesn't do it justice.* The serotonin in the intestines, at least as far as we know right now, doesn't seem to be linked to that in the brain, at least not in any functionally useful way. More to the point, to focus on one aspect of how one part of our body affects one function of our brain is to open the floodgates to every possible permutation of this kind of occurrence, and nobody has time for that. Just embrace the important point: the

---

* Superior science writer Ed Yong covers the crucial and complex role of intestinal bacteria in great detail in his book *I Contain Multitudes* (The Bodley Head, 2016) if you're interested in reading further.

things that influence our brain's ability to make us happy extend far beyond just our experiences and personal preferences.

Still, some persist in trying to find a simple solution for the conundrum of what makes people happy. The media have often run stories about certain equations and formulae that supposedly predict what makes people happy, what the happiest day of the year is, and the most depressing, and so on. Given all that's been said thus far about the complicated nature of happiness, it may seem surprising that it can be explained in a single equation or formula. And it *should* seem surprising, because it can't.

There are a number of reasons that these far-fetched formulae exist. One is something known as 'physics envy'.[31] Whatever you think of them, physics and maths are very 'fundamental' subjects, exploring the properties of numbers, particles, forces: basically the things that make up the uni-verse and our reality. These things typically obey complex but definable laws, meaning they behave in predictable and meas-urable ways in almost every context. As a result, as long as every variable is known, they can be defined with equations.

However, the more 'squishy' biologically based sciences, and psychology in particular, cannot really compare in terms of rigid laws and predictability. An object of a certain mass will accelerate at the same rate no matter where in the world you drop it, but the same person will behave and react in different ways depending on what room they're in, or who they're talk-ing to, or how recently they ate, or what they ate, and so on.

One result of this is that physics and maths are often thought of as 'proper' sciences. Academics and scholars in other fields, perhaps subconsciously, want to be taken as seri-ously as their peers studying physics, so try to copy physics and maths in their own fields, by producing equations for things as incredibly complex and messy as human behaviours, and moods. Like happiness.

So, bearing in mind all of the above, I knew the traps to watch out for if I wanted to study happiness. I knew what not to do. So, what *was* my task now? At this point, I'd done my research and, taking everything into account, had come up with a very carefully considered plan. I wanted to know where happiness comes from in the brain. To do this, I needed an MRI scanner to look at an active, happy brain. Because of the various issues around the use of human subjects unaccustomed to such studies, I'd figured the best option would be to use *my own* brain, given my background and experiences. So I needed:

1. To get access to an MRI scanner
2. To get inside it
3. To make myself happy (might need some pleasant stimulus or something, but if I'd made it this far with my plan odds are I'd be pretty happy anyway)
4. To have someone scan my brain
5. To look at the results to find which bit was most active, and is thus the source of happiness in the brain.

Simple. So now I needed to meet with a professor with the necessary resources and convince him to let me actually do this.

## Chambers of secrets

I arrived for my meeting with Professor Chambers at the pleasant Cardiff pub near his office where we'd agreed to have lunch. He was already sitting at the back of the room, and waved me a hello as I entered.

Professor Chris Chambers is a disarmingly laidback Australian in his late thirties. In what seemed to be a complete submission to cultural stereotypes, he was, at the time, wearing a T-shirt and baggy shorts (despite it raining outside). He is also completely bald, to a 'shiny' extent. I've met several younger male professors now who have little to no hair on their heads. My theory is that their big powerful brains generate so much heat that it scorches the follicles from the inside.

Anyway, I decided to take the plunge and just say what I wanted from him: 'Can I use one of your MRI scanners to scan my own brain while I'm happy, to see where happiness comes from in the brain?'

After about five minutes, he finally stopped laughing in my face. Even the most optimistic person would have to concede that this was not a good start. For the next hour or so, Professor Chambers explained to me, in detail, why my plan was ridiculous.

'That's not really how fMRI works, or how it should work. Back when fMRI was developed, back in the nineties, what we call the "bad old days" of neuroimaging, there was a lot of what we called "Blobology": putting people in scanners and hunting around for "blobs" of activity in the brain.

'One of my favourite examples of this is from one of the very first conferences I went to, there was a study being presented called "The fMRI of Chess vs Rest". Basically, you had people lying in a scanner, either playing chess, or doing nothing. The whole brain was active, but in different ways for the different scenarios, and in the chess scenario certain brain regions would show up as "more" active. From this, they then claimed these regions are responsible for the processes involved in chess. There was so much inverse inference applied: this part is active, and we do these things in chess, so that must be what those areas are for. It's working backwards. It's viewing the brain like a car engine; the idea that each brain region must do one thing and one thing only.

'This approach leads to these wrong conclusions; you see activity in a brain region and assign it a specific function. But it's completely wrong. Multiple functions are subsumed by multiple areas, which are handled by cognitive networks. It's very complicated. That's a problem with neuroimaging generally; it goes up a notch further when you're dealing with anything subjective, like happiness.'

Despite my openly joining in laughing at the naïve fools who thought you could use an fMRI to find out where chess playing comes from in the brain, I was dying of embarrassment on the inside. I'd hoped to do something very similar myself. I was, to utilise a term I'd only just discovered, being a total blobologist.

Turns out, it's one thing to use imaging tools to study something like vision; you can reliably control what your subjects see, and ensure each subject is presented with the same image to ensure consistency, and locate and study the visual cortex this way. But it's a lot trickier to study what Professor Chambers terms 'the interesting stuff'; the higher functions, such as emotions or self-control.

'The question is not "Where is happiness in the brain?" That's like asking "Where is the perception of the sound of a dog barking in the brain?" The better question is "How does the brain support happiness? What networks and processes are used to give rise to it?"'

Professor Chambers also touched on the issue I raised earlier: what *is* happiness, in the technical sense? 'What timescale are we talking about? Is it an immediate happiness, like "this pint is nice!"?* Or is it long-term and general, like your children making you happy, or working towards a goal, achieving contentment in life, being calm and relaxed, things like that? You have several levels of functioning in the brain supporting all this, and how do you unpack that?'

By now, I'd abandoned all hope of doing my half-cocked idea for an experiment, and admitted as much. Professor Chambers, despite my earlier fears about the ferocity of professors confronted by inferior intellects, was very nice about the whole thing, and said he would normally be willing to let me go ahead with it even if only to provide a useful demonstration of the technique. Unfortunately, fMRIs are incredibly expensive to run and several research groups are always vying for their use. It would probably upset a lot of people if he wasted precious scanner time allowing a buffoon to probe his own cortex for happiness.

I considered offering to pay the costs myself, but they were just too high. Not all writers are J. K. Rowling, and as generous as my publicist Sophie is when it comes to processing expenses submitted to the publishers, even they would baulk at a claim like this. £48 for a train ticket, £5 for a sandwich, £3 for a coffee, £13,000 for a day of fMRI. I couldn't see that slipping by the accounts department unnoticed.

Rather than just writing the meeting off as a lost cause, I decided to ask Professor Chambers if there were any other issues with the fMRI approach I should be wary of, before I attempted to rework my ideas to something more 'feasible'.

---

* We were in a pub so I'd bought us both a beer. By him mentioning it and me reporting it here, it now counts as a business expense so I can claim it back against tax.

It turned out Professor Chambers is a very keen and active individual when it comes to highlighting the issues and problems that afflict modern neuroimaging studies, and psychology in general. He's even written a book, *The Seven Deadly Sins of Psychology*,[32] all about how modern psychology could and should be improved.

There are several important issues about fMRI that clarified just how hard it would be for me to use it to set up an experiment to find happiness. Firstly, as stated, it's expensive. So studies that utilise it tend to be relatively small, using a limited number of subjects. This is an issue, because the fewer subjects you use, the less certain you can be that your results are significant. The greater the number of subjects used, the greater the 'statistical power'[33] of any results, and the more confident you can be that they're valid.

Consider rolling a dice. You roll it twenty times, and 25 per cent of those times you roll a six. That's five times you rolled a six. You might think that's a bit unlikely, but still perfectly feasible. It wouldn't seem significant. Now say if you rolled it 20,000 times, and 25 per cent of those times you rolled a six. That's rolling a six 5,000 times. Now *that* would seem weird. You'd probably conclude there's something up with the dice, it must be rigged or loaded in some way. It's the same with psychology experiments; getting the same effect or result in five people is interesting, but in 5,000 people it's possibly a major discovery.

Doing an experiment with one person, like I was hoping to do, is essentially pointless in the scientific sense. Good to know before I got started.

Professor Chambers then explained that this expense also means that very few experiments are repeated. The pressure on scientists to publish positive results (i.e. 'We found something!' as opposed to 'We tried to find something, but didn't!') is immense. These are more likely to be published in journals, to be read by peers and beyond, to improve career prospects and grant applications, and so on. But it's also best to repeat experiments where possible, to show that your result wasn't a fluke. Sadly, the pressure on scientists is to move on to the next study, make the next big discovery, so interesting results are often left unchallenged,[34] especially with fMRI.

So, even if I *could* run my experiment, I really should run it again and again, no matter what the result. Even if it was not giving me the data I wanted. And that's another thing.

The data produced by fMRI aren't nearly as clear as mainstream reports suggest. Firstly, we talk about which parts of the brain are 'active' during a study, but as Professor Chambers pointed out, 'This is effectively nonsense. *All* parts of the brain are active, all the time. That's how the brain works. The question is how much *more* active are these certain regions, and is it *significantly* more active than it usually is?'

To even get to the standards of 'blobology', you have to determine which blobs on the scanner are the 'relevant' ones. This is a big ask when doing something as fiddly as monitoring the activity of specific areas of the brain.* For starters, what counts as a 'significant' change in activity? If every part of the brain shows fluctuating activity all the time, how much does the activity have to increase by in order to

---

* And to make matters worse, fMRI doesn't even do that. The nature of how fMRI works, detecting how atoms scatter radio waves and all that, means it detects changes in blood oxygen levels in very specific parts of the brain. Brain tissue, like all tissue, uses oxygen whenever it has to do something, so a more active region will be using more oxygen, causing a greater change in the blood oxygen levels in that area, which can be detected by fMRI. It's still valid, but it's more of an indirect measurement of brain activity than you might have expected.

be considered relevant? What's the threshold it has to get to? This can vary from study to study. It's a bit like being at a pop concert of the latest megastar and attempting to work out who's the biggest fan by listening for the loudest infatuated screams; possible, but by no means easy, and a lot of work.

This, as Professor Chambers explained, results in another glaring issue.

'fMRI has a huge what we call "Researcher degrees of freedom" problem. People often don't decide how they're going to analyse their data, or sometimes even which question they're going to ask, until after they've run their study. And they go ahead, and they explore, and they have this "garden of forking paths" problem, where in even the simplest of fMRI studies there are thousands of analytical decisions to make, each one of which will slightly change the outcome they get. So what researchers will do is mine their data at the end to find a result which is useful.'

This comes about because there are many different ways to analyse complex data, and one combination of approaches may provide a useful result, where others wouldn't. It may sound dishonest, somewhat like firing a machine gun at a wall then drawing a target around where the most bullet holes are clustered and claiming to be a good shot. It's not *that* bad, but it's heading that way. But then when your career and success depends on hitting the target and this option is available, why wouldn't you do it?

But this was just the tip of the iceberg regarding all the issues that come with running fMRI experiments. Professor Chambers had potential answers and solutions to all of these problems: reporting methods of analysis in advance of actually doing them; pooling data and subjects between groups to increase validity and bring down costs; changing the way scientists are judged and assessed when awarding grants and opportunities.

All good, valid solutions. None of which helped *me*. I came to this meeting hoping to use some high-tech wizardry to locate where happiness was coming from in my brain. Instead, my brain was left reeling with the myriad problems of advanced science, and feeling distinctly unhappy about it.

Professor Chambers eventually headed back to work, and I made my disappointed way home, my head buzzing with more than just the two beers I'd consumed during our talk. I'd started out thinking it would be relatively easy to determine what makes us happy, and where happiness comes from. It turned out that even if the scientific techniques I'd hoped to use were straightforward (which they really aren't), it had become obvious that happiness, something everybody experiences, everybody wants, and everybody feels they understand, is far more complicated than I'd anticipated.

I see it like a burger. Everyone knows what burgers are. Everyone understands burgers. But where do burgers come from? The obvious answer would be 'McDonald's'. Or 'Burger King'. Or another eatery of your choice. Simple.

Except burgers don't just pop out of the void fully formed in a fast-food restaurant's kitchen. You've got the beef (assuming it's a beef burger) that's been ground down and formed into patties by the supplier, who gets the beef from a slaughterhouse, which gets it from a livestock supplier, who raises cattle on grazing land and rears them and feeds them, which consumes considerable resources.

Burgers also come in buns. These come from a different supplier, a baker of some description, who needs flour and yeast and many other raw materials (perhaps even sesame seeds to sprinkle on top) to be pounded together and placed into an oven, which needs constant fuel to burn and create the necessary baking heat. And don't forget the sauce (extensive quantities of tomato, spices, sugars, packaging assembled by industrial-level processes) and garnish (fields dedicated to growing vegetables, which need harvesting, transporting and storing, via complex infrastructure).

And all these things just provide the basic elements of a burger. You still need someone to assemble and cook it. This is done by actual humans who need to be fed, watered, educated and paid. And the restaurant supplying the burgers needs power, water, heat, maintenance, etc. in order to function. *All* of this, the endless flow of resources and labour that your average person doesn't even register, goes into putting a burger onto a plate in front of you, which you might eat, absent-minded, while staring at your phone.

A convoluted and complex metaphor perhaps, but that's the point. Looking closely, it seems that a burger and happiness are both familiar-but-pleasant end results of a ridiculously complicated web of resources, processes and actions. If you want to understand the whole, you must also look at the parts it's made up of.

So, if I wanted to know how happiness worked, I needed to look at the various things that make us happy, and figure out why. So, I resolved to do just that. Right after I'd had a burger.

Don't know why, but I was suddenly craving one.

## 2

# There's No Place Like Home

There are many ways to describe my mood as I left my meeting with Professor Chambers, but 'happy' wasn't one of them. It was a long and miserable walk home as I worried about what to do next.

However, as I arrived at my street, something odd happened; I caught sight of my house, and I started to feel better. No intense euphoria or giddy high, but a definite sense of positivity and relief, probably not something I'd usually notice but in my current downcast state it was a stark change. When I actually went into my house my mood improved further. My thinking changed from 'What am I supposed to do now?' to 'What AM I going to do now? What'll I look at next?' The former suggests despondency; the latter action, motivation, engagement. Arriving home had improved my mood. A lot of people report similar experiences, the sheer relief and pleasure in getting home after an arduous journey or long work day. It's a very common feeling. You could say our homes, in various ways, make us happy.

Is that right? Is it just the sense of an otherwise unpleasant task or series of events coming to an end when we arrive at our own doorstep, or is there something about our home that triggers positive feelings in our brain? What is going on in there?

Neuroscientifically, the idea of our homes making us happy doesn't make much sense. Neurologically, we quickly adapt to familiar things. Neurons stop responding to signals and stimuli that occur repeatedly and predictably.[1] Think about when you walk into a kitchen where someone is cooking something pungent, like fish. It stinks! But after a few minutes, you stop noticing. Then someone else comes in, moans about the smell, and you wonder what they're complaining about. That's habituation. When you get dressed you quickly stop 'sensing' the feeling of your clothes. That's habituation too. Studies show that people can even get used to electric shocks,[2] as long as they're predictable, and relatively mild. Habituation is a powerful process which means the brain immediately focuses on any sudden changes in our situation, but if it hangs around and doesn't do anything important, the brain essentially loses interest.

We spend maybe half our waking (and nearly all of our sleeping) lives in our homes, so you'd think they'd be the last thing our brain responds to. Why, then, does the home provoke a response in our brains that results in us being happy?

As with 'happiness chemicals', this argument seriously oversimplifies how the brain works. Our brains and nervous systems do stop responding to things, as long as they are not *biologically relevant*. This is key; it means we stop responding to things that have no *biological consequence*.

We need food. We eat it several times a day. But do you ever get 'bored' of food? You can tire of *types* of food; eat nothing but pasta for a week and you'll quickly get fed up with it. But the act of *eating*, consuming food, that never gets dull.* The most mundane meal can, when hungry, provide feelings of satisfaction, contentment, pleasure, happiness? Even a glass of water seems like divine ambrosia from God's

---

* Notwithstanding extreme situations such as eating disorders, of course.

own keg if you're hot or thirsty, because it's *biologically relevant*. Our brains recognise it as something we need in order to stay alive, so rewards us with pleasurable feelings when we obtain it.[3]

It's not just nice things. People may quickly get used to the temperature of water that they're immersed in, but not if it's literally scalding, because this causes severe pain, something our brains rarely, if ever, fully adapt to. The initial intensity may subside, but pain suggests damage has occurred, or *is* occurring, to the body. This is *very* biologically relevant, so mustn't be ignored. Pain even has its own dedicated neurotransmitters, receptors and neurons,[4] all dedicated to 'nociception', the perception of pain. It's essential, if unpleasant.

Our brains 'overrule' habituation when it comes to important things. And if they're positive, beneficial things, this activates the reward pathway, meaning we experience some form of pleasure whenever we encounter them. So, there are some things we remain keen on and responsive to, regardless of familiarity.

What's this got to do with home? Are our homes 'biologically significant'? Quite possibly. Consider all the essential things that happen within your home: nourishment, sleeping facilities, warmth, even plumbing (expelling bodily waste is another vital function).

Pavlov's famous dogs learned that an innocuous sound meant that food was forthcoming and responded enthusiastically as a result,[*5] establishing the fundamentals of

---

* This enthusiastic response was quantified by measuring the amount of saliva they produced, a reliable indicator of the expectation of food in dogs. Yes, one of the most famous psychology experiments in history involved the collecting and measuring of dog drool. Nobody ever said science was all glamour.

associative learning, where mental connections are made between separate occurrences. It takes the formidable human brain no time to learn that our home is where all our biological essentials can be found, so we form a positive association with it.

But this is a *learned* thing. Our home isn't doing anything biologically relevant, it is just where biologically relevant things happen. Is there anything to suggest that our brains respond to our homes directly? To answer this, look at the fact that homes are *naturally occurring*.

Homes aren't something humans invented for somewhere convenient to keep our shoes and iPads. They occur everywhere in the natural world, in many different forms: birds' nests, anthills, termite mounds, rabbit warrens, bears' dens, and many more. Countless species have homes; we humans are just the first to come up with doorbells.

If something is common in a wide range of species, it strongly suggests a biological need is being met. Evidence points to a sense of *safety*. Biologically relevant things typically keep us alive, help ensure our survival. But, in nature, it's not just lack of food that kills you; there are countless dangers and threats out there, most obviously predators, but environmental hazards too. Abundant food is useless if you slip into a shadowy ravine and break your neck.

As a result, even the most basic mammal has evolved a complex and sensitive threat detection mechanism. In humans, regions like the amygdala, subgenual anterior cingulate cortex, superior temporal gyrus, fusiform gyrus and more[6] form part of an intricate network that rapidly processes information from the senses, evaluates it for anything that looks like a threat, and triggers the appropriate reaction (e.g.

the fight-or-flight response). This threat detection system is incredibly useful when exploring new, unfamiliar locations, looking for resources or mates but not knowing if hungry carnivores are lurking in the shadows.

But it's not something we just switch on when we think it might be useful, like taking an umbrella when it might rain; it's ever-present, ready to spring into action at the faintest whiff of danger. Some evidence suggests it can even be triggered by a simple shape. A 2009 study by Christine Larson and colleagues[7] showed these threat-detecting areas became more active when presented with basic 2D shapes composed of downward-pointing 'V's. Essentially, pointy triangles set off the threat-detection system. Not substantially so, or we'd be quaking in fear at the sight of the alphabet, or kites. But still. It even makes a certain sense; many natural dangers, like a wolf's face, fangs, talons, spikes, etc. all have general 'V' shapes. Our evolving brains spotted this, and became wary of it.

Our brain's threat-detection systems are sensitive and persistent, but constant fear and paranoia is very debilitating, as anyone with chronic anxiety will tell you.[8] It's an extremely stressful way to live, impacting negatively on the health of your body and brain. People who suffer from anxiety often feel unable to leave home. This makes sense; familiar places are less dangerous – you've been there often and haven't died, so your threat-detection system isn't dialled up to maximum, like a sniper on amphetamines. Instead it's turned way down, like a night watchman at a village shoe shop; still vigilant, but not really anticipating having to do anything. It's very helpful to have access to a reliably safe and familiar place, to stop the effects of constant fear and stress. And *voilà*, a clear biological benefit of having a home.

What is interesting is that when we are at home, we can more easily focus on anything out of the ordinary. If you're in an unfamiliar restaurant and hear a glass smash, it's distracting for a moment, but then technically *everything* is, because everything's unfamiliar, so you pay it only cursory attention. You're in your own home and hear a glass smash? That's unusual, suddenly you're primed for danger because you know it's a hazard (doubly so if you're home alone). Some studies even suggest we can detect and recognise threatening stimuli faster in a familiar environment than in an unfamiliar one.[9] It makes sense; there's less to distract us, our brain is used to 'ignoring' the environment around us, so anything that differs from that gets our attention much faster. A tiger in the jungle can be hard to spot; a tiger on a cricket pitch is not.

Providing a reliable environment where we can be stress-free is one way that our homes can make us happy. This isn't to say that our home is automatically stress-free. It can be a source of great anxiety, but more often than not this is down to unfortunate-but-fixable problems (rising damp, a broken boiler, etc.) or the people you share your home with, like an abusive partner. One study from the eighties even suggests that having their adult children living at home is a potent source of stress for elderly couples if the parent–child relationship is antagonistic,[10] something that's no doubt becoming more of an issue with today's chaotic property markets. Nonetheless, on balance our homes themselves are usually a means of reducing stress, not increasing it.

The provision of a safe environment has a further important consequence: sleep, another essential function we mostly do in our homes. Sleep and mood are known to be linked in complex and potent ways. Disrupted or limited sleep can cause irritability, stress and low mood in humans,[11] so just by allowing us to get sufficient sleep, our homes increase the odds of us being happy. Scientists have even studied what happens when people try to sleep in unfamiliar places such as hotels. In a 1966 study, electrodes were strapped to the heads of dozens of volunteers who spent four nights sleeping in a lab,[12] while their brain activity was recorded via electroencephalograms (EEG). The study found that volunteers' sleep was much reduced and disrupted for the first night, but not for subsequent nights. This was the first recorded demonstration of the phenomenon known, unsurprisingly, as the 'first night effect'[13], where people struggle to sleep as soundly as normal when in a new location. It could be a five-star hotel, in a four-poster bed on pillows stuffed with angel feathers, but you still won't sleep as soundly as on your own dented mattress, because in unfamiliar surroundings, part of our brain remains slightly more 'awake', keeping us vigilant at a subconscious level.*

Thus far, we've focused on 'home' as the actual physical structure we inhabit. But people can have home communities, home towns, even home countries. While the latter may be more of an abstract appreciation (a whole country is far too vast for a single human to have any tangible connection with it), the human sense of home clearly doesn't end at the four walls we reside in.

This applies to other species too; you never see an elephant in a nest (although that would be amazing) but while

---

* Some species take this to extremes. An obvious example is unihemispheric sleep[14], where one half (hemisphere) of the brain sleeps while the other stays awake and keeps the body doing whatever it needs to. This occurs in dolphins, while swimming with their pod, and migratory birds, while flying over oceans.[15]

they might not have a specific dwelling it doesn't mean they don't have a home. Many animals, including elephants, have a 'home range', a specific area they move within but rarely beyond. Others have 'territory', and while animals with a home range don't necessarily mind others sharing it (or might just avoid them), animals with a territory will defend that area against intruders. A moose seeing another moose in its home range will do little more than grunt at it, but a tiger in another tiger's territory gets bloody, quickly.[16]

We humans can and do live in places with population densities that would drive other species out of their tiny minds. That we don't object to sharing our environment with others suggests that humans have a 'home range'. Then again, people who've been burgled say the most upsetting aspect is the sense of violation, the knowledge that someone unknown has been in your home without your awareness or permission. It is not uncommon for people to be suspicious, unfriendly or downright hostile to strangers or anyone different in 'their' neighbourhood. Maybe humans have a mix of home range *and* territory, whereas most animals stick to one or the other? Either way, it shows our brains are keenly aware of our homes, even if they cover an impressively wide area.

That this is even possible is because of our spatial awareness, a system that means we know where we are and where we're going at any given time. The hippocampus and surrounding brain areas in the temporal lobe appear to be key for spatial mapping, navigation, orientation and other similarly important abilities.[17] The hippocampus is crucial for long-term memory formation, so this makes sense; in order to know where you are, you need to remember where you've been.

That's not all. Certain neurons in these areas respond to things like 'head direction', activating only when the head is pointing a certain way, allowing the brain to keep track of which way it's going. Others are 'place cells', which only activate when you're in a specific recognisable spatial location,[18] allowing the brain to keep track of familiar places, like pins in a map. There are also 'grid cells', which seemingly provide an awareness of our position in space. If you get up and travel across a room with your eyes closed, you're still aware of where you are and where you moved. Grid cells are believed to be necessary for this ability.[19]*

There are even 'boundary cells', neurons that activate when we come to a specific environmental boundary, like a river that marks the end of your territory, or a door that leads from your house to outside. They're neurons which activate whenever our senses detect where our current environment 'ends', letting us know that we're about to cross an important threshold. Most of these cell types are found in the hippocampus or associated regions nearby.

These sophisticated spatial systems let us know where we are and where we're going, especially with regards to our homes. 'This is my home.' 'This is where my home ends.' 'My home is in this direction.' This helps explain why we can often find our way home without even thinking about it, like when we're somewhat intoxicated.**

This is relevant because it leads to another important

---

* In truth, grid cells performing this sort of function have been discovered in animals like rats and monkeys, but not quite pinned down in humans yet. We may use them like any other species, or we may have evolved a more diverse, flexible system. It's impressive in any case.
** Known in the UK as the 'beer taxi', where you wake up in bed the morning after a drunken night with no memory of how you got there.

role of a home, and helps resolve what may seem like a contradiction. If our brains need a home because it provides comforting familiarity, as unfamiliar environments trigger the threat-detection system, how do we explain things like curiosity? Rats, mice, cats (despite the old proverb) and many other creatures also show 'novelty preference', a spontaneous interest in things they haven't encountered before;[20] curiosity by another name. Yet these same creatures also demonstrate neophobia, a reflexive fear or anxiety when presented with anything unfamiliar.[21] How can novelty preference and neophobia co-exist in the same brain? Well, situations change; a useful response in one scenario would be inadvisable in another. Applauding someone after a wedding speech is appropriate, after a eulogy might not be. Even at the more basic levels, a properly working brain takes the current context into account when deciding how to respond.

Some studies show that mice don't necessarily get stressed by novel/unfamiliar things themselves, they're more scared of *not being able to get away from them*. If you allow a mouse to access an unfamiliar place from a familiar one, it'll cheerfully explore said new place and things within it. But put in a strange place with no escape, they show fear and anxiety.[22, 23] Apparently, unfamiliar things are only scary if you encounter them when you've nowhere safe to retreat to. Another crucial role for a home; it provides a safe environment from which to explore, investigate the new, find useful resources, and survive.

For the human equivalent of this kind of anxiety, consider homesickness. One theory is that the distressing feelings of missing home evolved to discourage vulnerable humans from wandering off alone and away from the safety of the community.[24] The brain clocks when we're cut off from our homes, and the reaction is often negative. One particularly stark example of this is the occurrence of 'cultural bereavement',[25] something that affects migrants when they become aware of the loss of the social support, customs and cultural norms of their home country, to an extent that damages their mental wellbeing. This is felt particularly keenly by refugees, who invariably have to establish a new home in a new country in the wake of already very traumatic circumstances. Mental-health issues in these populations are relatively high as a result.[26] Perhaps this is the most obvious example yet of a link between our homes and our brains that impacts on our happiness.

In summary, our homes are biologically relevant; they provide safety and security, things that are essential for our survival and general wellbeing, so our brains respond positively to them, making us happy. It's a nice, neat system isn't it? So, let's throw several spanners into the works.

The typical first-world human can expect to have many different homes over their lifetime. Some they remember fondly, others they barely remember at all. I had multiple homes during my university years, but can't even recall their addresses today. In contrast, I can recall my family home in painstaking detail. Why this discrepancy? If the point of a home is just to satisfy a biological requirement, why are some homes desirable, when others are not? Also, why does anyone ever move house? As well as giving up a knowingly safe environment in favour of a less certain new one, moving house is genuinely one of the most stressful experiences (outside of traumas or calamities) a typical human can go through.[27] The time, effort, cost, uncertainty, loss of personal control, all trigger the stress response in our brains. Why would we willingly put ourselves through such an upheaval?* It is an especially odd decision when you consider the phenomenon of risk aversion;[28] a cognitive bias inherent in most brains which plays down potential benefits and exaggerates potential losses presented by a decision. Many of us stick with the familiar; perhaps we always order the same thing on the menu whenever we eat out: 'The other thing *might* be better, but I *know* I like the thing I usually have, so I won't risk it.' If our brains steer us away from changing our meal choices, why would it not stop us from changing our home?

It didn't make much sense to me, so I decided to speak to someone who has extensive experience in the world of property buying and renting, in a place that has proven to be an immensely popular location for people to set up home, New York City.** As it happens, back in 2016 I had the unexpected and surreal pleasure of featuring in Brick Underground, a website that bills itself as the 'daily survival guide for buying, selling, renting and living in New York'. So, I asked if they'd like to return the favour, and ended up talking to Lucy Cohen Blatter, Brick Underground writer and journalist, and someone who has interviewed hundreds of home owners and sellers/vendors in the proverbial Big Apple.

---

* Obviously, plenty of people have changed homes for circumstances beyond their control, such as financial issues, disasters, work, and so on. Here, I'm specifically referring to people who do so of their own free will.
** This is also true of London, somewhere much closer and easier for me to investigate. But as a British person who doesn't actually live there, I'm sick of hearing about London! So, forget it.

## Start spreading the news

Lucy is a born-and-bred New Yorker. If anyone knew the intricacies and appeal of the city, she would. Also, her husband is from Birmingham in the UK, so she has experience with British accents, which was helpful as my own dense Welsh accent and rapid delivery means non-Brits often find me 'a challenge'. I called Lucy at her New York apartment and started by asking why New York was such a popular place for people to want to set up home.

'I think the reasons are two fold: probably the most obvious factor is work. If you want to find work, New York is where you can find it. You've a much greater chance of finding a job here than maybe anywhere else.'

This made sense, but how relevant is it here? People move home because of their jobs and careers all the time.* But this doesn't really tell us about what they want from a home, as it's a secondary consideration to getting (and keeping) a job. People choosing a home because it'll help them get a job are like people choosing which airline to fly on holiday with; while helpful, the flying isn't the most important bit. Nobody likes airline food *that* much. This is a pretty big area to cover though, so I opted to come back to the whole 'work' thing later and, for now, press on with homes.

'The other main thing about New York is the variety of culture,' Lucy continued. 'As well as all the different communities and influences, if you wanted to you could do something

---

* Especially scientists. Someone pursuing a career in highly competitive research typically has to go wherever there's funding and jobs available in their field, and this is rarely within commuting distance of where they currently live.

different every night, there are just so many choices and options for entertainment, exploring, socialising, and so on. Also, New York, particularly after the events of 9/11, has a strong sense of community, of solidarity. There's a palpable "energy" in the city that you don't get in other places.'

The people around us, the community we inhabit, have a direct and significant impact on our moods and our thinking. Our brains, essentially. This isn't specific to New York but let's just accept at this point that our interactions with other people have a big effect on our happiness. Again, more on this later.

However, that New York offers such a multiplicity of options and entertainments *is* worth looking at more closely. Why would the dizzying variety of shows, films, exhibitions, etc. prove intoxicating to so many?

Basically, never underestimate the importance of novelty. Even though it can sometimes be scary, as we saw with neophobia, it's often a potent and rewarding quality. Many animal studies have shown that environmental enrichment (putting things in the surrounding to make it more complex and interesting) has tangible, beneficial effects on the brain, up to and including increased brain growth, enhanced hippocampal development likely boosting memory and related processes,[29] and even prevention of seizures and neuronal death.[30] Maybe living in a buzzing city like New York, Helsinki or Berlin is actually very *good* for us, at least in terms of our brain's functioning? Maybe this is why so many writers, artists and other creative types flock to NYC?[31] As well as the huge cultural scene in which to promote and sell their creations, could it be that just living in a stimulating place inspires the creative thought processes?

Novelty is also important when it comes to pleasure. Specific areas of our brain, namely the substantia nigra and ventral tegmental area (both near the very centre of the brain), show increased activity when we're presented with novel stimulation. But these regions are also more active when we're anticipating a reward. Most importantly, certain parts show increased activity when we're anticipating a reward *in a novel context.*[32] Essentially, novelty can heighten the reward response.

What's this mean, in plain English? Well, it strongly suggests that, assuming they're all pleasurable, our brains find new things more enjoyable than familiar ones. A joke is never as funny the second time you hear it. Your first kiss is the one you specifically remember, not the countless subsequent ones over the course of your marriage. Novelty causes our brain to enhance the positive stimulation we experience. Novelty, it could be said, makes us happier.

This occurs via activity in the reward pathway discussed earlier. Studies show that if we have fewer dopamine receptors in the reward pathway, we need more stimulation in order to experience 'normal' levels of pleasure[33] and pursue thrills to a degree most others would find excessive. We end up, in short, doing things like bungee jumping, excessive drinking and living in New York, perhaps.

The endless novelty offered by city life is another likely reason why a home there may be desirable for someone with a standard (or not) human brain. It may also help to answer our question about why anyone would do something stressful when we're programmed to be risk averse. There are many complex theories and baffling mathematical models about the exact properties of risk aversion, but one important conclusion is that aversion to risk is overcome *if the potential reward is big enough*. Sitting in a tin can strapped to thousands of tons of liquid explosive is a huge risk, but astronauts do it willingly because they get to go into space, a reward considered 'worth it'. And for many people, risking losing your safe, familiar home for a less certain one elsewhere, like New York, is considered worth it for the potential rewards. Better jobs, more to do, more people to meet, are all clear and tangible benefits that their existing home can't provide. Remember, it's not just the building we inhabit but the surrounding area that our brains consider 'home'. As a result, our brains weigh up the possible pros and cons, and while usually it exaggerates the cons, sometimes the pros are heavy enough to tip the balance; a safe, familiar home in the middle of nowhere can rank lower than a potentially inferior home in the middle of *somewhere*.


## Headspace


Studies suggest a sense of continuity is necessary for a place to feel like a home,[34] meaning we're less likely to feel at home in a place that we know we'll be leaving relatively soon. This is why, if you have multiple addresses in a place like New York over a short time, the city itself can feel more like your home than any of the individual structures you lived in, as while the buildings themselves might not have provided a sense of continuity, the wider city has.

But still, nobody wants to live in a dump. You seldom get people seeking a home accepting the very first place they see; they invariably look around, for the 'best' available option. So, there must be certain qualities and aspects of a home structure that people seek out, perhaps something that our brains actively respond to, even if on a subconscious level. I asked Lucy about the specific nature of the homes people seek out in New York.

'Often space is the main concern. It's a very densely populated city, so anywhere that has enough space for you is going
to be very in-demand. In fact, one of the main reasons people
end up leaving New York is because they need more space.
They want to start a family or get a bigger place for other reasons, and they often can't do that in the city, so have to move
elsewhere.' This was interesting in itself, but what Lucy told
me next suggested that, in New York, space is more valuable
than money.

'I did this series of articles where I looked at the homes of
the extremely rich, the one-percenters, and even they had to
sacrifice something to live in the city centre. In Manhattan,
they lived in two-bedroom apartments, extremely big and
nice two-bedroom apartments, of course, but these people
could afford mansions elsewhere. Yet they're willing to live in
smaller places in order to have a home in a desirable location
like New York.'

It seems we all have to make sacrifices to buy the space we
want. But *why* is space so important? If a home simply satisfies
an instinctive desire for safety, we shouldn't need excessive
space. If anything, as long as a home is big enough to contain
your possessions and all the essentials (plumbing, a bed, etc.),
the less space the better. Smaller homes are cheaper to heat,
easier to maintain, easier to keep secure, and so on.

But a small home means you can't acquire any new stuff,
have friends over or expand your family. There's also the
question of social status; having a big home is a sign of wealth
and success.

Our desire for space goes deeper than these concerns;
our brains *need* a specific amount of space to feel calm and
to avoid feeling stressed. There's a whole field dedicated to

the study of our sense of space called proxemics. It was first developed by anthropologist Edward T. Hall in 1966.[35] He suggested that a typical person has four 'zones' of space, with surprisingly clearly defined boundaries; intimate space, personal space, social space, public space, each of which stretches progressively further away from our body.

More recent data suggests that people's sense of space varies from person to person and culture to culture.[36] People have different ideas of what counts as 'near' or 'far', for instance, and one study suggests that people who have an exaggerated sense of proximity are more likely to suffer from claustrophobia.[37] But even without clinical problems, people have a keen sensitivity to space; as we've already seen, several aspects of our brains' sensory systems have been directly linked to the processing and encoding of our 3D environment. If we're in a restricted space, our brain is aware of this on multiple levels. And it doesn't like it. If a stressed person yells, 'I NEED SOME SPACE!' and storms off, they are probably speaking literally.

The overall point of this is that, given how our brains process space, a very small home is likely to be less tolerable. Restricted space means entrapment, inability to know what's happening nearby (beyond the close walls), reduced options for escape. Our home is meant to be the place we can retreat to when we're stressed or anxious, but if it's too small our brain's threat-detection system remains active, which is exactly what our homes are meant to prevent. Also, if we're already stressed or anxious, some evidence suggests our personal boundaries 'expand', meaning we are less tolerant of people and things being too close.[38, 39] And thus, practical and architectural concerns aside, some homes are too small in the *psychological* sense. It's not that we *can't* live in smaller homes, it's just more difficult to feel positive, to be happy, in them.

And there's another major factor that means people want spacious homes: privacy.

Most people don't live alone. This is a good thing for the most part – sharing social interactions is important for happiness, as stated. But, does anybody want to be around people *all the time*? Even the most outgoing, enthusiastic extrovert needs down-time in their own private space, even if it's just to sleep. Interacting with others, however pleasant, gives the brain work to do. Social psychologists acknowledge that in practically any context there comes a point where the interaction becomes irritating for all involved,[40] basically because everyone gets mentally fatigued. As a result, we eventually need to withdraw and avoid interactions for a while. This allows individuals both to 'recharge' and to avoid social upsets and all of this helps to maintain important relationships, giving rise to clichés like 'man caves'. Sometimes we need to be around people; sometimes we need our privacy.

This ability to retreat is no doubt particularly important for city dwellers, who exist in an intense high-stress environment where there are people absolutely everywhere. A home there puts you right in amongst the action, but also allows you to get away from it, meaning you can engage with things on your own terms, providing a sense of control and independence, which is something people tend to like. Yet another way your home facilitates your happiness.

Speaking of space, homes with gardens or some green space nearby are always in high demand. Lucy pointed out that the most coveted (and expensive) streets in New York are those with a view over Central Park, and homes with gardens are often considered superior to those without, even if said garden is little more than a scrap of land. Obviously being able to go outdoors but stay within the 'confines' of your home provides an excellent sense of space, and we know that's important. But even those of us without gardens tend to fill our homes with pot plants, window boxes and the like. What is this compulsion to surround ourselves with greenery all about?

It isn't purely an aesthetic preference; interactions with nature and biodiversity seem to have tangible positive effects on our brains. One explanation for this was given by Stephen Kaplan, in what he dubbed Attention Restoration Theory.[41] Kaplan argues that our brain's attention system is usually 'active', darting about constantly, dealing with a barrage of people and being consciously directed to focus on the currently most important thing (e.g. the book you're reading). This takes effort, energy, and is therefore taxing for our brain. If after a long frantic day, where your attention has been constantly in demand, you've ever felt the need to 'veg out' and do something utterly mindless like work your way through an Adam Sandler box set, then you know this to be true.

Natural surroundings engage our attention passively, a process Kaplan dubs 'fascination'. Our attention is able to wander more in natural surroundings and the brain gets a break from directed attention, which requires neurological effort. Our brains can rest and recuperate, replenish resources, strengthen connections, enhance cognitive faculties and improve our mood. To this end, Kaplan dubs environments full of greenery and biodiversity 'restorative spaces'.

The beneficial properties of green spaces can't be stressed enough; they even affect our bodies. One study reported that hospital patients with very similar conditions recovered faster if they were in a room with views of trees and nature, as opposed to views of brick walls.[42, 43] You can see the evolutionary logic of this. As we saw earlier, unfamiliar environments trigger our brain's threat-detection system, so technically an environment with nothing in should be more relaxing than one full of unfamiliar things. But an empty environment is, in nature, a barren one. For a creature hoping to survive, that's no good. Rich, green environments, full of the resources essential for life; those are where you'll find what you need. A creature with a brain that responds positively to these environments, that's drawn to them but isn't overly stressed by them, has a definite survival advantage. It would be immensely impractical if everything unfamiliar made us fearful.

## MY home, MY castle

'I know many couples who are from different backgrounds where, say, the wife grew up in a rural area, the husband in a city. And they live in a small apartment in New York, and they have to keep their bike in the living room. The wife will see this and think, "There's a bike in the living room, we need more space, we have to move." The husband genuinely won't see anything wrong with this,' Lucy told me. She continued: 'You need to be a certain type of person to live in New York. Some people are "New York" people; they really get on in the city and love it here. Others just aren't, they don't last, and quickly leave. A person who loves it here but has to leave for financial reasons or other things beyond their control, they're still a New York person.'

What Lucy was getting at was individual personality is a significant factor in how and why our homes make us happy. This wasn't a surprise, but I wasn't pleased to hear it nonetheless. Trying to explain personality in terms of neurological processes is like trying to figure out the anatomy of a cat by dissecting the hairball it's coughed up; there's undeniably a link, but good Lord is it convoluted.

Personality and individual differences, to some extent, risk undermining much of my argument so far. For example, I've said the human brain prefers spacious homes, but this isn't universally true; at the time of writing there's something of a new fashion for tiny homes emerging.[44] Similarly, I've argued that our brains *need* privacy, but some people opt to live in alternative communities (cults, communes, etc.) where this is scarce at best, and some of the most densely populated cities in the world offer little chance for space and privacy.

Attempting to address this tricky conundrum was why, an hour after I said goodbye to Lucy, I was in the car, heading west up the M4 motorway. While it may sound like I was 'doing a runner', I was actually going home. That is, my first home, the one I grew up in.

My reasoning went as follows: maybe our personalities are a big factor in why our homes make us happy, and why some are better than others, but we don't emerge from the womb with a fully formed personality. So where *do* we get it? Well, if we invoke the classic nature vs nurture debate, in the case of personalities it seems to be a clear mix of both. Our genes play a major part, but so too do our experiences as we grow and develop. Things our parents do and say, our interactions with our peers, and, crucially, the environment we inhabit; these all play a part in determining our personality as adults.

And what environment do we spend most of our time in, particularly as we're growing up? Our *home*!

The idea that our childhood homes help shape our personalities in later life isn't just an assumption on my part; there's evidence to support it. An extensive study by Shigehiro Oishi and Ulrich Schimmack, published in 2010, interviewed over 7,000 people who had moved homes frequently as children (for instance those whose parents were in the military and were regularly posted to new places). Their findings showed that there was a direct link between changing homes often during childhood and reduced psychological wellbeing, life satisfaction and meaningful relationships as adults.[46] Put simply, growing up without a stable and consistent home as a child can make you less happy as an adult. A very neat link between our home, our brain, and happiness.

Listening to Lucy's explanation about the appeal of New York, I found it interesting that in spite of gaining a better understanding of why so many people choose to move there, at no point did I ever find *myself* wanting to do so. I've never seen the appeal of living in a busy city. I shy away from large crowds, don't deal well with constant background noise, find very tall buildings quite intimidating and oppressive, and I'm not naturally assertive enough to overcome these foibles. I suspect this is due to my growing up, my brain developing, in a small, isolated, quiet, rural community, where I never learned to appreciate, or even just tolerate, such things.

Technically, then, if my guess was right, I should feel happiest in my childhood home where I acquired all these traits. So, I thought, hell, why not test this assumption? Let's go to my childhood home, and see what that does to me.

I didn't grow up in a 'house'; I grew up in a pub. The Royal

Hotel, Pontycymer. We moved there when I was two, and maybe that experience, going from a tiny house with just my parents to such a relatively huge building filled with inebriated strangers, was excessively jarring for my tiny, developing brain. Perhaps this is what imbued me with my dislike of crowds and noise, and my reluctance to draw attention by asserting myself. This would explain why I spent my formative years wandering about a busy drinking establishment, trying to avoid eye contact with the patrons and helping myself to crisps (I was a shy and, eventually, overweight child).

So, off I drove to the pub. For science!

I admit I felt apprehensive en route. We moved out of the pub when I was around fifteen, and I had been back as a customer just once, when I reached legal drinking age (well, close enough). But it had felt very weird, seeing other people occupying what was your home for years. I imagine it was like attending the wedding of someone you were previously in a long-term relationship with; a confusing emotional mix of nostalgia, affection, regret, envy, anger, bitterness, and more.

It had been close to twenty years since I'd last visited. How would I feel? Would the intervening years colour my sense of 'coming home'? Would the memories have lost their potency, meaning I'd now view my childhood home as one views an old garment; something once useful but no longer needed? Or, given how environmental cues are known to trigger related memories,[47, 48] coupled with the fact that childhood memories tend to be the most vivid,[49] would visiting my earliest home cause me to revert somewhat to the person I was at the time, as some claim can happen?

In the end, all this musing was for naught, as there was no sign of life in the pub that was my childhood home. It was a mess of boarded-up windows and doors, broken glass, and hideously overgrown brambles. Derelict, and clearly had been for some time. The economic decline that has been hitting my home region since the 1980s had claimed another victim.

How was I feeling at this point? Well, 'happy' certainly didn't describe it, but I struggled to find a word that did. It was just very, very strange. Triggered by the presence of my old home, my brain was regurgitating a whole host of fond memories: playing soldiers on the outside stone stairs, go-karting in the sloping beer garden, chasing the family dog after he'd snuck into the kitchen and stolen an unguarded steak, decorating the whole place every Christmas, and so many more.

Simultaneously, as I peered in through gaps in crumbling windows, all of these cherished reminiscences were being overlaid in real-time by images of these fondly remembered places riddled with decay, rot, ruin, destruction and obscene graffiti. It was, to say the least, an uncomfortable experience, trying to process two such disparate emotional reactions at once. Imagine if, after the aforementioned wedding of your former lover, you're watching the honeymoon car drive away and it suddenly explodes.

I sat down on the pavement outside my former front door (after brushing aside the shattered glass fragments) and, like a dedicated neuroscientist, tried to figure out a neurological explanation for my response. Logically, it was just a building, one that I've no tangible connection to any more. The fact that it was now in a state of disrepair was undeniably a shame, but it didn't really have any bearing on me and my life now.

Except, it did! I was experiencing a powerful emotional reaction to the state of my old home, which I wouldn't be if it wasn't deeply relevant to me in some way. Clearly there was more going on in my brain than a simple abstract recognition of a former dwelling. It felt like a part of me had died. Was that fair? If our home interacts with our brain in the many different ways I've described, and our brain is 'us', is it a stretch to argue that our homes form part of our identity? Looking into it further, it seems that no, it isn't.

Professor Karen Lollar, of the Metropolitan State College of Denver, once lost her home in a fire, and wrote a paper about her experiences.[50] It was, as you can imagine, deeply traumatic. Homesickness can be very upsetting, as we've seen, so how much worse must it be to lose your home altogether? Even psychological associations recognise this,[51] particularly in the event of a home being physically destroyed, as with a house fire or natural disaster. There's no specific term for this, but comparisons can be made to cultural bereavement, in that a sudden involuntary loss of your established home in traumatic circumstances is bound to be psychologically damaging.

As Professor Lollar eloquently states in her paper, 'My house is not "just a thing" . . . The house is not merely a possession or a structure of unfeeling walls. It is an extension of my physical body and my sense of self that reflects who I was, am, and want to be.'

Scanning studies indicate that this 'my home is a part of me' notion is reflected in our brain's workings. One study showed increased activity in the medial prefrontal cortex when subjects viewed objects they thought of as 'theirs', in comparison to those belonging to someone else.[52] More interestingly, this same brain area showed increased activity when subjects considered adjectives and words they felt described their personality. In a nutshell, the same brain regions that process our sense of self and personality, are also used in recognising our possessions and property. While this study looked at individual possessions rather than homes per se, our home is our biggest, most prominent possession. We spend the most money on it, adapt it to our tastes most extensively, and we keep most, if not all, of our other possessions in it.

This can also expand beyond the limits of the basic structure that is our home. There's a psychological theory known as 'place identity', where a person has attached such meaning and significance to a certain place that it contributes to and influences their sense of self.[53] This is mediated by a sense of place attachment,[54] which is where an individual forms a strong emotional bond to a specific place. Have you ever been looking for a new home, walked into a potential property and thought 'yes, this is the one', before you've even really looked around? Or gone to visit a place and instantly fallen in love with it? So much so, you keep coming back, or relocate there as soon as you can? A friend of mine, Chris, once made plans to spend a few months travelling the world, starting in Japan. Nearly a decade later, he still hasn't left there yet. Sometimes a place just ticks all the boxes you never realised your brain had, and you immediately *identify* with it. That's place identity.

So, what conclusions could I draw from all of this? From my own studies, the insights provided by Lucy about the appeal of New York, and my own unexpectedly traumatic visit to my childhood home, what did I learn about how our homes and our brains interact, and how this does or doesn't make us happy?

It seems we humans are compelled to find a home because it satisfies an innate drive for safety and security, which allows our brains to stop scanning for threats and dangers and being constantly alert. Our brains also quickly learn that our homes are where other biological essentials are addressed, like food, warmth and sleep. In removing a range of immediate stresses, home induces a positive association in the brain, contributing to our wellbeing, our happiness.

But some homes are preferred over others. Homes that offer greater safety and security, that offer more space, more privacy and access to greener environments. Our brains prize these homes over others. Our brains are also sophisticated enough that our sense of home isn't limited to the structure we inhabit, but the surrounding area too, and if a home is in an area that means greater access to novelty, stimulation and opportunities then these will usually be preferable to those that don't.

On a more cognitively complex level, much of what we seek and like about a home is determined by our underlying personality and preferences. It is not just a useful object or possession, it is a major part of our lives. We spend so much time in it and so much of our energies on it, acquire so many memories and associations with it, that our brains literally recognise it as an extension of ourselves, our identity. Again, this can be applied over a wider area, with people incorporating a place or location into their sense of identity (e.g. 'I'm a New Yorker').

There's a major point to consider here; a home may make you happy, but the location and nature of your home is typically determined by other factors, like work, money, variety, safety, proximity to friends and family, and many other things. The most popular homes always have something else to offer, they're not just 'a pleasant place to live'. Sometimes, they're not even that. This just emphasises that, as important as they are, our home isn't the *most* important thing for our happiness.

Taking all this into account, maybe the interaction between our brain and our home is *too* fundamental. Perhaps trying to explain how your home makes you happy is like trying to explain how your legs make you happy; there are many ways in which they can, but that's not what they're really *for*. It goes deeper than that.

Perhaps it's fairer to say our homes help us to avoid unhappiness, rather than making us happy. This may sound like splitting hairs, but the two aren't quite the same thing; 'not being in debt' is different from 'being rich'. Our homes are too involved in every aspect of our lives, and our brains interact with them in too many ways, to definitively say they make us happy in one particular way. Maybe, then, the point of a home is that it satisfies a sufficient number of our basic needs and requirements so that we are then able to focus our energies on other things that make us happy? Work, entertainment, family, relationships, creativity, etc. Rather than making us happy, our homes make it easier for us to be happy. If there's a conclusion to be had from all this, maybe that's it?

I started this chapter with a purpose; to show that our homes make us happy, and that there is a clear neurological explanation for why this is. I end it sitting amidst broken glass in front of the ruin of my childhood home. I know I'm one for elaborate metaphors, but even I found this rather excessive.

# 3

# Working on the Brain

As I navigated the chancy valley roads on my way home from where I grew up, it was clear my investigation into the mechanisms of happiness wasn't going great. Plans to use neuroimaging had been crushed by a shiny-headed professor, and considering the links between home and happiness had brought me the shattered remains of my childhood domicile. Hopefully, you'd forgive me for feeling unhappy about this.

Except you wouldn't have to, because weirdly, I didn't. Maybe it was progressing poorly, but I was still getting to write an actual book, a dream for many. It's all a matter of perspective, so I was still enthusiastic and upbeat, and keen to see where my investigations would take me to next.

In February 2017, 'next' turned out to be Bologna, Italy. Home of, among other things, the oldest university in Europe,[1] and an airport that sells Lamborghinis in the arrivals hall. As you do.

I was there to do a talk for the MAST foundation, a cultural and philanthropic institution that's part science museum, part gallery, part restaurant, part nursery, part university, part gymnasium, and probably more – like a space colony accidentally built in an ancient Italian city. While there I got to see their latest art installation, which proved surprisingly useful.

It was composed of videos of various real-life scenarios: a bare-chested, middle-aged Mediterranean man guiding a digger cutting slabs from a huge marble quarry;[2] young Ghanaian males searching for valuable scrap amid mounds of technological waste from Western countries; office employees stamping endless streams of documents; German factory workers finishing vehicles rolling off automated production lines, and so on.

The theme of the exhibition was 'Work in Motion', showing the many ways in which people around the world work. Given that we spend huge chunks of our adult lives working, with some estimates calculating that we spend a solid decade doing it,[3] the nature of our work will inevitably have lasting effects on us, one of which is how happy, or unhappy, we are. Hardly a controversial claim; a bad or unpleasant job will make you utterly miserable, and work-related stress is a major problem.[4] And on the other side of the coin, we've all met that person who 'can't wait to get out of bed in the morning', because they just love their job.* So, it seems clear: good jobs make us happy, bad jobs make us stressed and miserable.

Except, this is the brain we're talking about here; when has it *ever* been so straightforward?

As the MAST exhibition revealed, the work people do varies immensely, and the average Westerner typically has around twelve different jobs before they're fifty – a figure that seems to be rising.[5]** But no matter the job, it's always a human being and brain doing it. So, what *does* working do to our brains, that makes us happy or unhappy?

---

* Typically said with such cheerful smugness that you have to resist the urge to strangle them with the hilarious novelty tie they're invariably wearing.
** Meaning the average number of jobs. Someone's age is always rising, that's just how time works.

## Working hard on the brain

The most basic definition of work is 'energy and effort expended in performance of a task'. Essentially, all jobs and tasks require you to spend energy and effort in some form, be it physical or mental. But even at this crude and simple level, work has an appreciable effect on the brain that can, and probably does, make us happier.

Copious evidence shows the more physically active you are, the better your brain works.[6] Makes sense; the brain, a biological organ, needs energy and nutrients (more than other organs).[7] Increased physical activity strengthens and improves the heart, reduces fat and cholesterol, speeds up metabolism, all of which improves the supply of blood and nutrients to the brain, increasing its ability to do . . . anything, really.

Physical activity seems to have an even more 'direct' effect on the brain, by increasing output of Brain Derived Neurotrophic Factor, BDNF, a protein that stimulates growth and production of new brain cells.[8] This could explain the many reported neurological benefits of physical activity, such as enhancing learning ability and memory,[9] increased hippocampal volume and higher levels of grey matter throughout the brain.[10] Studies also suggest that children who engage in more physical, sporting activity often do better on academic tests.[*][11]

So if our work compels us to engage in physical activity, the positive effects this has on the brain could well make us

---

[*] However, don't force your child onto the football field just yet. Does sporting activity make children smarter? Or are smarter, determined children good at both sports and academic assessments? As with most things neuroscientific, it's not clear-cut.

happier. Boosting our learning and related faculties makes us smarter (arguably), and despite the term 'blissful ignorance', evidence suggests that greater intelligence makes you (slightly) happier.[12] Also, physical exercise releases endorphins,[13] the 'happiness chemical' discussed in Chapter One. And, of course, improved general physical health means we have greater capacity to do things that make us happy, as we aren't held back by poorer health and stress resulting from no exercise.

Likewise, mental activity also has apparent benefits for brain and body – good news for those of us whose jobs don't involve any more physical exertion than is required to arrive at the office on time. Higher levels of education have been shown to be reasonable protection against dementia and Alzheimer's,[14] to the extent that autopsies have revealed that highly educated people had brains that were seriously degraded by the disease, but had shown no obvious clinical signs of it prior to death.[15] The overall conclusion is that a more active brain is also a hardier brain.

We know the brain is flexible and adaptive, constantly forming new connections and reinforcing existing ones, as well as letting unnecessary ones fade away. The brain operates something of a use-it-or-lose-it policy, so the more used a brain is, the more connections and grey matter it'll have in place. Age and entropy will take their toll, of course, but the more well-used brains are able to withstand it better. They have a larger 'cognitive reserve', as it's often labelled. The more we use our brains, the cleverer we become. Sort of.

So, working means we engage in some form of physical and/ or mental activity, and this (eventually) improves our brain's functioning, making us smarter and happier. How handy!

Just one slight problems with this conclusion: it's nonsense. Sure, physical activity, effortful work, and improved brain functioning and happiness are linked, but that's clearly not the whole story. You staple a person to a horse and they are very much linked, but it doesn't automatically follow that centaurs exist. The real explanation is far more complicated and difficult, like trying to staple a person to a horse.

For example, if physical exertion makes us happier, why do we regularly avoid it? Why aren't we all constantly jogging to the quarry for a nineteen-hour shift digging up rock with our bare hands? If physical effort is automatically cheering, the Ghanaians who spend all day crawling over mountains of jagged metallic waste should be happier than pampered corporate executives sitting at a desk in their own spacious high-rise office. A dubious conclusion, to say the least.

The reality is, while physical exertion may have useful benefits, it can quickly become damaging and painful, hence 'forced labour' is scientifically recognised as a form of brutal punishment, not a treat. Basic physics means engaging in such activity, such 'work', requires energy. Our bodies are good at using and storing energy, but it's obviously *finite*; we can't just keep going and going like a mechanical battery-advertising rabbit. Too much physical activity means our energy reserves are depleted and our bodies are damaged.

This has obvious implications for survival. We've discussed how the brain links actions with rewards to encourage us to do them. But what if the action is *too* demanding? A jungle cat spending a whole day hunting a tiny shrew, *Tom and Jerry* style, will have spent far more energy in the pursuit than it gains from consuming such small prey, so it's *lost* energy overall. Such behaviour, repeated often, will literally kill it. In human terms, imagine a job that pays reliably and regularly, but your daily wage is less than what it costs to travel there each day. The fact that you're rewarded for your efforts isn't the point; it's that you're not rewarded *enough*.

Thankfully, our brains have seemingly evolved to prevent this. A study by Irma T. Kurniawan and colleagues in 2013[16] analysed subjects being made to exert large or small amounts of effort in order to gain or avoid losing money, and found evidence of a neurological system which anticipates the need to exert effort, based in the anterior cingulate gyrus and dorsal striatum. These regions show increased activity when subjects are aware that greater amounts of effort are required, coupled with raised activity in the ventral striatum when the reward experienced was greater than anticipated, but perhaps most interesting is the reported finding that this effect was reduced when the amount of effort exerted was higher.

To translate: it seems these brain regions automatically assess how much effort a task requires, what the outcome is and, crucially, combines them both to decide *is it worth it*? If you've ever looked at a job in front of you and thought to yourself, 'You know what, I can't be bothered,' well, now you know why. However, this effort evaluation system, much like the threat-detection system, never stops – even when it's not really needed, which is all kinds of ironic. This has major implications for our work, and beyond.

If this neurological system just determined whether effort is warranted and left it at that, it might be OK. But the truth is, it makes our brains so sensitive to wasted effort as to actively try to prevent it, even going so far as to alter our thinking and behaviour. For instance, in a study led by Nobuhiro Hagura at UCL[17] subjects were asked whether a cloud of dots on a screen was moving left or right, and then had to push a corresponding handle to answer. One of the handles was made increasingly hard to move, meaning it required more effort to report the dots as moving in a certain direction. Alarmingly, the results implied the subjects *stopped seeing* the movement that cost more effort to report, even when it was definitely there.

Consider the implications: *our very perception* is altered to avoid unnecessary effort. Our consciousness, our view of the world, is subtly changed to avoid pointless hard work.

In truth, evidence suggests the brain does this worryingly often. People will reliably say an odour labelled something like 'spring meadow' smells much nicer than one labelled something like 'used toilet water', even if both odours are identical.[18] Objects relevant to our current goals can appear 'bigger' in our vision, and hills and climbs can seem steeper than they are if we're in a negative frame of mind, like if we're afraid of heights, or if we're aware that the climb will be arduous because we're carrying a heavy load. It seems our perception is often altered to discourage us from things the brain's decided it doesn't approve of.

One explanation argues that unpleasant things incur a related emotional state (disgust, frustration, etc.). The brain has to basically 'create' a representation of everything we perceive based on the raw data it's receiving from the senses. This obviously involves a lot of extrapolation and calculation, but our emotional state is offered to the brain as a sort of 'short cut' when doing this. Say we're stood at the edge of a cliff, the brain essentially says, 'Well, I could use all the relevant visual signals to work out how high up I am, but I'm currently experiencing nervousness and fear, so I must be obscenely high, so I'll say that.' This, apparently, twists our perception of things. Tiny spiders seem gigantic to arachno-phobes, other cars seem terrifyingly fast to learner drivers, and if you hate your job your workplace may look grey, miserable and depressing, even if the casual observer would disagree. What we perceive isn't based purely on a detailed analysis of the sensory information supplied, but is tweaked and altered by the emotional associations triggered by whatever it is we're looking at, and wasted effort seems to be very good at triggering emotional associations.

So, to summarise; our brains don't like it when we put effort into something for no obvious benefit, and when our brains decide we don't like something, engaging with it inspires negative feelings and perceptions. Even more succinctly, doing work for no obvious gain makes us unhappy! If you spend hours assembling a flatpack wardrobe only to have it collapse as soon as you put a single sock in it, your reaction can range from crushing despair to teeth-gnashing rage, but certainly not happiness.

Think of how often this happens in the world of work. You spend months working on a grant proposal or project, only for it to be rejected. You do your best day in, day out, but are repeatedly passed over for promotion. You deal politely with customers only for them to be abusive and rude. Years of output are rendered pointless when your company undergoes a merger. The very nature of much of modern employment means a sense of futility is easy to come by, so perhaps it's no wonder that many (if not most) people refer to work in a generally negative way, and regularly wake with a sense of at best apathy or at worst dread on Monday mornings.

Expending physical effort may be good for our brains and have a positive effect on cognition and happiness overall, but it's a slow and subtle process. In comparison, investing effort for no obvious reward is a sure-fire way for our brains to label a task as unpleasant. And because of the nature of many modern jobs, effort going unrewarded is a very common occurrence.

So, if this is the case, why do we work at all?

## Work is not its own reward

Here's a fact about me: I used to have a job embalming and dissecting dead bodies for a local medical school. They were used to teach students about surgery and anatomy. Since then, I always 'win' any debate about who's had the worst job. But it's a pyrrhic victory, admittedly.

As unpleasant and unsettling as this job was, though, I did it for nearly *two years*. Perhaps my experience is more grim than most people's, but this isn't an uncommon phenomenon. Many people complain constantly about their awful jobs but still drag themselves to the workplace every day and do what's expected of them, loathing every minute. Why? *How?*

The obvious answer is, because they *must*. We may have created a frighteningly complex world around us, but humans still require essentials like food, water and shelter. But now we don't go out and find these things ourselves; we buy them. With money. And we get money by working. So surely, it's wrong to say that our efforts at work aren't rewarded, because we're paid for them?

Technically, yes. The brain does recognise money as a valid reward for our efforts, at a fundamental level. Evidence indicates that financial reimbursement provokes a response in parts of the brain like the mesolimbic reward pathway[19] that are also stimulated by biologically significant rewards (food, sex, etc.). So, getting money makes us feel good. A rat or a pigeon won't feel the same about money; they'll just see a handful of metal discs or colourful paper, worth a cursory sniff maybe, but little more. We, however, can grasp the inherent value and importance of money, and that working is how we obtain it.

The importance of money can't be overstated. There's a reason the question is often 'what do you do for a *living*?'; not having enough money is genuinely a threat to our survival, which explains why Western psychologists rank losing your job in the top ten most stressful things you can experience.[20] Lack of money also triggers the brain's ever-sensitive threat-detection system. Working is the most obvious, risk-free, socially acceptable way of preventing this. So, as well as providing a reward for our efforts, money also provides a sense of safety,[21] hence the term 'financial security'.

It's no wonder, then, that we spend so much time working in jobs we detest, despite how much our brains may object to doing so. It also hints at how our work can make us happy, at least partially; as with our homes, satisfying basic needs and providing a sense of safety typically prompts a positive response in our brain. This also explains why, as we saw in the last chapter, our jobs often determine where we choose to live; we need money for a home, and a job for money.

It's not *just* about the money though, because as we know the brain habituates to anything that becomes reliable and familiar enough. Your first pay packet can make you very happy; a psychological burden (worrying about paying your bills) has been lifted, and you now have more choice and more financial freedom to do things. But after weeks or months of the same amount of money arriving in your bank at the same time, you become desensitised to it. It's just that something becoming predictable loses 'potency', hence finding £50 in your old trousers feels better than getting your usual £500 pay.

Thankfully, there are other aspects of our work that our brains recognise as rewarding, because our brains aren't solely concerned with satisfying basic organic needs. Some scientists differentiate between survival needs and 'psychological' needs,[22] which are things that aren't strictly essential for our biological survival, but that we find fulfilling for more cognitively sophisticated reasons. One of these is a sense of control.

In the 1960s, psychologist Julian Rotter developed the concept of the locus of control.[23] If you think you are responsible for what happens to you, you're said to have an internal locus of control. If you believe you're at the mercy of others and external events, you have an external locus of control. Several studies have linked an internal locus of control to higher levels of wellbeing and happiness, even health, in groups as diverse as college students[24] and elderly war veterans.[25] Makes sense; if you control events, then you can prevent bad things from happening. If you don't, there's little you can do to prevent the bad things. Which sounds more stressful?*

Some argue that locus of control is an inherent trait, something essentially 'fixed', but there's evidence that it's more a

---

* This isn't always the case, of course. For some, a sense of control means they're far more stressed and feel personally responsible when things go wrong, while by contrast, feeling one has no control over anything means no pressure. Humans, as always, vary quite a lot.

learned thing, and can be changed via our experiences.[26] The neurological mechanisms are unclear, but at least one study links locus of control, along with self-esteem and responses to stress, to the size of the hippocampus,[27] suggesting that experience and memory are indeed key factors. But then, other evidence suggests that a sensitivity to feelings of control and an aversion to losing it forms at a very young age, even before we're able to walk![28, 29] It's no wonder that infants really hate the word 'no'.

Whatever the underlying mechanism, the implications for our work are obvious; if we have a job with authority and responsibility, we're more likely to perceive a sense of control, which our brain likes, so we end up happier.

Your work can reward you with a sense of control, but it can also provide a loss or lack of control, which can be psychologically harmful, sometimes even clinically so.[30] Jobs that strip you of autonomy with strict rules/policies (dress codes, micromanagement, etc.) and/or make you constantly beholden to others (telesales, retail, etc.) are widely regarded as unpleasant and a source of stress. It may be that businesses insisting 'the customer is always right' has actually had a very damaging effect on their workforce.

Related to control is competence: our ability to do something and do it well. The brain's ability to accurately assess our performance and abilities is a crucial cognitive function. It allows us to make valid decisions about what we should and shouldn't do. You're walking down the street and see someone collapse; you DO get your phone out and call an ambulance, because you know you're capable of this. You DON'T try to perform open-heart surgery on the pavement using your car-keys and a ballpoint pen, because you know that's beyond you and would cause considerable harm. Exactly *how* the brain judges its/our own performance is uncertain. There is evidence linking the tissue density, the amount of important grey matter packed in, of the right ventromedial prefrontal cortex, in the frontal lobe, to accuracy of self-appraisal,[31] so presumably that area plays a role. But in any case, our brains seem to place a lot of value on competence.

Our jobs give us ample opportunity to acquire and demonstrate competence; if you can't achieve a minimum level of competence in your job then you usually lose it, and given how the brain recognises our work as important for our survival, the desire for competence is bound to be high. It also ties into our brain's effort-evaluating system, as doing something we're not competent at is considerably more effort than something we're an expert at. Driving to the shops to pick up milk is a mundane chore for many, but for those who can't drive or don't know where the shops are, it requires a herculean effort. Clearly, our competence is an important facet of our brain's underlying calculations.

This can even be demonstrated in the very structure of our brains. Experienced London taxi drivers have been shown to have enlarged regions of the hippocampus, specifically the regions dedicated to complex spatial navigation,[32] and musicians proficient in instruments like piano or violin have been shown to have significantly larger areas of the motor cortex dedicated to fine hand and finger movements.[33] Our jobs essentially compel us to perform actions and behaviours repeatedly, which means our brains have time to adapt to them, making us far more proficient at them. And this can make us happier, because the brain likes being competent.

Also, many jobs offer a variety of ways to *measure* our

competence. Sales targets, bonuses, promotions, performance reviews, pay grades, employee-of-the-month awards – these are all things which provide a reasonably quick and definitive measure of how 'good' someone is at their job. Our brains do seem to like measuring things, and appear to have specific regions dedicated to doing so. A 2006 study by Castelli, Glaser and Butterworth[34] suggested that the intraparietal sulcus, part of the brain's parietal lobe, is integral to the brain's processing of measurements, and that it even has separate systems for specific, numerical measurements (e.g. 'There are thirty-eight chips on my plate') and more 'analogue', relative measurements (e.g. 'There are more chips on his plate than mine, I am never eating here again'). The intraparietal sulcus has also been regularly implicated as having a fundamental role in integrating information supplied by the senses and linking it to our motor systems, and other facets that control our behaviour,[35] so this all adds up.

And yes, pun intended.

So, for various reasons, our brains desire a sense of competence, and when we feel we're competent, we're more likely to be happy. Our work offers us greater opportunities to improve our competence, and to have this competence objectively confirmed, which is nice. (Unless of course, our competence is criticised, which is not.)

Work also offers other types of reward, such as exposure to novel things and situations (something the brain likes, as the previous chapter revealed, and that explains why jobs that are crushingly repetitive are often described so negatively) and greater opportunities to interact with other people and make social connections (covered later). The take-home point here is that, while most people work because they need the money, the brain's mechanisms offer several other ways in which work can reward us and satisfy instinctive needs and desires, potentially making us happy – even if your job involves dissecting cadavers.

## Where your brain sees itself in five years' time

All this research about how the brain deals with working eventually dredged up a memory of a job-related incident from my own life, one that made me physically cringe. So, counterintuitively, I'm going to share it with you.

I once, upon starting a new job, attended a compulsory all-day 'introductory seminar', where new employees were given presentations about the company's aims and objectives and so on. It was, predictably, crushingly dull, and during an afternoon session on company values, I fell asleep in my chair.

I was startled awake when the speaker said, 'So Dean, what are our three core values?' Thanks to my semi-conscious and baffled brain, I answered with, 'Um, serve the public trust, protect the innocent, and uphold the law?' There followed a long, awkward pause, undoubtedly caused by the fact that I'd supplied the prime directives of the eighties sci-fi character Robocop, rather than the company core values we'd only just been told about. Not exactly relevant in this context.

Leaving out the ridiculous film trivia element, my experience is a common one. You hear many tales of people struggling to remain conscious during yet another tedious workshop, seminar or conference that their employer nonetheless insists upon. It's seemingly just a fact of modern working life, much to the annoyance of everyone outside of upper management. Why? Why insist on this rigmarole when it just annoys everyone and distracts from their actual jobs?

Here's the thing: many businesses and organisations *want* their employees to be happy, and go to great (and costly) lengths trying to achieve this, via away days, team-building exercises, motivational consultants and seminars, retention schemes, feedback surveys, workplace perks and much more. And while some may do this purely out of generosity of spirit or concern for their workforce, the cynical truth is that happy employees are more profitable.

There's compelling evidence to suggest that happier employees are up to 37 per cent more productive. So if you have 100 employees and make them all happy, they could be doing the work of 137 people, for no extra cost. Conversely, unhappy employees can be 10 per cent *less* productive.[36] Add in other things like happy people being healthier[37] and less likely to complain, then of course businesses are going to want to make their employees happy, even if they consider them nothing but worthless peons.

Unfortunately, making diverse groups of individuals happy *on command* is mind-bogglingly difficult, unless you take drastic measures like putting ecstasy in the water coolers (Lord knows what THAT would do for productivity). This is because the brain keeps throwing up hurdles and complications, an important one of which is motivation (something else they try to entice in employees) and how it works in the neurological sense.

Motivation is largely directed by goals; we have a goal we want to achieve, and our motivation directs our behaviour accordingly to get to it.[38] For most creatures, this is a simple arrangement; the goal is 'get food' so they're motivated to hunt and forage, or the goal is 'don't die a brutal, messy death' so they're motivated to avoid the huge creature with a distressing array of teeth.[39] But we humans, with our infuriating intelligence, have taken these basic motivational processes and woven a vast array of complex behaviours out of them. People previously assumed we were simply motivated to do things we liked, that made us happy, and avoid anything we didn't like. Freud himself argued this with Freud's Hedonic Principle.[40] But humans and their brains aren't that simple.

Even at the day-to-day level, how the brain deals with motivation makes it tricky to exploit. Common sense would suggest that most employees are motivated to earn money. So, offer them more money, they should be more motivated, right? Wrong! There's evidence to suggest that, in certain situations, paying people more can make them *less* motivated to do something. Why would this happen?

Well, motivation can also be classed as extrinsic and intrinsic. Extrinsic is when you do things for external rewards, intrinsic is when you do them for internal ones, because you find them personally enjoyable or satisfying, or they conform to your personal drives and ambitions.[41] Wanting to be a doctor because you want to help people and do good in the world, that's intrinsic motivation. But if it's because you want decent pay and job security, that's extrinsic motivation, because those are things supplied by outside agencies.

Importantly, intrinsic motivation seems the more potent kind, because, you could argue, the rewards come from within our own brains.[42] The contradiction produced here is that sometimes if you coerce people into doing something via rewards like financial incentives, they feel less like it's their *decision* to do it, so their motivation becomes contingent on said rewards. Basically, once the reward is received/removed, the associated motivation fades away. This doesn't seem to happen if it stems from an internal, personal source, if it's our own decision to do it.

One study focused on children who were given art supplies to play with. Some were given a reward if they used them, others were just allowed to do their own thing. Later, they were given the art supplies again, with no additional instructions. Those who hadn't been rewarded, who'd played with them voluntarily, showed greater interest and enthusiasm in doing so again.[43] From this, employers may conclude that their employees would be happier and more motivated if given more autonomy and control over their jobs, rather than simply more pay. Maybe chefs in chain restaurants would be happier with their work if they could serve dishes how they wanted, rather than a manner determined by 'head office'? Or maybe, I don't know, writers of science books would be willing to give their advance back in return for the ability to set their own deadlines?*

However, a word of caution. Firstly, it's not an either/or thing; people still want, *need* to be paid. The study with children and art supplies may seem compelling, but children don't have mortgages and children of their own to provide for. And we also know now that the brain is predisposed to putting in minimal effort, so given the freedom to do 'their own thing', many people would likely not do much at all. Many employers clearly know this, as almost every job has numerous rules and regulations that workers must adhere to if they want to keep

---

* Note to my editor: no, they wouldn't.

said job. But then, such rules reduce employee autonomy, making workers unhappier and less productive. It's a tricky balancing act, with no obvious solution to this.

However, there's a whole other level of motivation that we need to consider. While most animals live 'in the moment', our human brains can and do think further ahead. This means that as well as immediate goals, we also have long-term or 'life' goals. Ambitions, basically. There's evidence suggesting that life goals that you work towards make you happier and more content than focusing solely on the basic needs of survival.[44] Robert Agnew's General Strain Theory of Criminology[45] even cites failure to achieve goals as a major category in the causes of criminal behaviour. Clearly, having long-term goals and ambitions has a significant effect on happiness and behaviour.

Why? Well, many psychologists argue that our brains hold separate images of ourselves; our 'ideal self' and our 'ought self'.[46] Our ideal self represents a goal, an ideal state we want to embody *eventually*, and our ought self is the behaviour we feel we need to demonstrate *now* in order to achieve it. Our ideal self may be a champion athlete, at the peak of physical fitness. Our ought self would therefore be someone who goes to the gym and avoids eating pizza and cake, because that's what you 'ought' to do to achieve your goal. Evidence suggests that our ideal self is a big factor in our happiness while working.[47] Basically, if our brain recognises that what we're doing moves us closer to achieving our ideal self, we're happier. If it doesn't, we're not. So if you're doing a job that doesn't conform to your own personal goals, or even actively detracts from them, it's harder to be happy while doing that job.

The ideal scenario, therefore, would be to work in a job that you actively want to do and that facilitates your own life goals, so that your ambitions and those of the people you work for complement each other. Many employers seem to be aware of this, at least on some level. One explanation for the constant efforts to engage employees and make them share the 'company vision' is that the higher-ups want to explain their plans and intentions in order to convince employees to share these goals. Hence the common interview question, 'Where do you see yourself in five years' time?' If an applicant says, 'in an assistant manager position in the procurement department', this suggests they'll be a dedicated and enthusiastic worker for the company. If they say, 'I hope to be an Olympic gold-medal-winning tap dancer', chances are they won't be 100 per cent invested in the job.

Unfortunately, in today's world, getting a job that aligns with your life goals is far from certain. Many children express a desire to be an astronaut, but few long to be baristas. But which are you more likely to meet in your daily life? Nothing *wrong* with being a barista, of course, but making a double-shot grande soy latte° surely can't compare to flying a spaceship, and there's nothing any coffee shop manager can do to change that. And having a job unrelated to your goals – particularly if it's a high-pressure, demanding one – can be directly detrimental to achieving them. Work-related stress is already very mentally draining, meaning you have less enthusiasm and willpower to pursue things beyond routine behaviours,[48] so you resort to bad habits and stress-relieving indulgences (e.g. binge eating, alcohol), which impact on your health, and move you further from your ideal self. So, further

---

° I'm no coffee enthusiast so am not sure if this is a real thing. Sounds right, though.

stress, and unhappiness, will follow. Is it any wonder so many people complain bitterly about their jobs?

Bizarrely, it's still possible to be happy while doing a job you didn't necessarily plan to do, because in certain circumstances your brain will flip and decide you do actually want to do it. Say you want to be an Olympic gold-medal-winning tap dancer, and just work in an office to pay the bills, to allow you to follow your dreams. But then you realise you'll never achieve this goal (maybe someone cruelly points out there's no such thing). Now, this presents your brain with a problem. Up to now, there's been a valid (sort of) rationale for working in an office, something you don't otherwise want to do. But now that's gone. Now you're doing something you don't want to do, for no good reason. This sets up an inconsistency in the brain, and it doesn't like this. So it needs to resolve this.

One thing it can do is accept that you've failed, that all your efforts have been in vain, that you're not competent, that you should leave your job and start again. This may seem like the logical, sensible option, but it comes at considerable psychological cost. Or, it could just change what you think, so now it turns out you *do* want to work in an office. All that other stuff was just a childish pipe dream. This is proper work. If you knuckle down and focus on your career, you could be assistant manager of the procurement department in five years!

This is a form of cognitive dissonance,[49] which is where, faced with incompatibility or inconsistency between our thoughts, behaviours and actions, our brains will do what's necessary to resolve the conflict. And if it can't change the reality, it'll change what you believe and think. And thus, your ideal self, your life goals, are changed, because the brain instinctively protects us from stress and failure where possible, even when it's not exactly logical to do so. So, while our work may not be of any use when it comes to pursuing our ambitions, there are situations where our brain will instinctively alter our ambitions so it *is*, to increase our chances of being happy.

However, this is just one possible outcome. The harsh truth is, because almost every workplace conforms to a hierarchical structure (that's just how humans do things[50]), the odds are stacked against being happy at work. The brain craves control, and working for someone else limits that.[*] Also, despite regular accusations to the contrary, the bosses and other higher-ups are typical humans too, and their brains mean they have the same inherent wants and needs as their underlings. Unfortunately, the motivations of the typical worker (e.g. get as much money as possible for doing as little work as possible) aren't compatible with those of the bigwigs responsible for the success of the company (e.g. get employees to work as much as possible for as little money as possible). Is it really any wonder that the seeming majority of people regard their work as a 'necessary evil', something essential but not exactly celebrated? No wonder the work–life balance has become such a familiar concept.

## If I were a rich man

At this point, I felt I had a good idea of how our work affects our brains, and how this determines our happiness. Physical

---

[*] Being self-employed should logically solve that, but this introduces a great deal more uncertainty, and the fact that your income depends on the whim of customers and clients, so it doesn't really.

work can improve brain functioning in tangible ways, but unre-warded work is something the brain seems evolved to avoid as much as possible. We work because we need money in order to survive, which our brains recognise on a fundamental level, meaning people still persevere with jobs they hate. But our brains instil in us other needs and desires – to be in control, to be competent, to be looked up to – and our work can give us these things, or deny them, which also affects how happy we are. We humans can also have long-term ambitions, and whether our work helps or hinders in achieving these is another big factor to consider.

Looking at this altogether, you could say it's *all* about goals. Psychologists and neuroscientists often speak in terms of goal-directed behaviour,[51] which describes pretty much every action and behaviour that isn't purely habitual or reflexive, because every conscious action has a purpose, a goal, that prompts it, and the brain seems to have an array of complex systems that facilitate this.[52] Survival, financial stability, control, competence, approval – these are technically all goals, and, along with any overarching 'life' ambition, help explain why we work and the effects it has on us. So that explains how work affects our happiness, and why we all do it.

Except . . . it doesn't entirely. The obvious issue here is, goals can be *achieved*. That's sort of the point. 'Follow your dreams' may be a trite or corny saying, but some people do actually succeed in making them come true. It's entirely possible to have total control and complete financial security, be the best in your field, achieve your ambitions, and all that, sometimes at a surprisingly young age. What then? Do people just . . . stop?

Apparently not. Consider all the super-wealthy business leaders or sporting champions who carry on despite winning everything. They have more money and respect than they could ever hope for, so if they don't *need* to work, why do they? What about it makes them happy?

I wanted to find out, but how? If I wanted to be scientific about it, I'd take a regular person, remove all their reasons for working, and analyse how it affects them. Sadly, asking my publishers for a million pounds 'so I can give it to a stranger and see what happens' was met with a weary sigh and a curt refusal. So, next best thing again, I'd talk to someone already in the relevant situation. Basically, I needed to talk to a working millionaire.

Where to even find one, though? Should I just wander the bars and clubs in London's affluent Mayfair district, seeing if anyone fancied a chat? But then I realised there was one who came to my first book launch: property developer, entrepreneur, businesses consultant and fellow Welshman Kevin Green. Maybe he'd be willing to help?

As well as fitting the bill, a quick bit of research revealed that in 1999, after winning a Nuffield Scholarship,[53] Kevin studied the attitudes and personalities of high achievers, interviewing Bill Gates, Sir Richard Branson, and others of that ilk. Kevin Green wasn't just a convenient person to talk to, he was an *ideal* person to talk to. So, I went to visit him.

We know that money is the most obvious reason for working. In fact, Kevin also provides training and coaching seminars about how to make money, a predictably popular and much sought-after service. But people don't inevitably pursue jobs and careers based solely on the likely financial rewards. So I asked, from Kevin's perspective, how big a factor is money when it comes to people's working lives and happiness?

'My angle is, if you're choosing what line of work you're going to go into, you've got to be very passionate about it, and enjoy it. I think if you chase money, money runs away from you. Some people get very wealthy, then they lose it, because they've just been chasing money.'

An interesting perspective to hear from a millionaire entrepreneur maybe, but one with supporting evidence. A 2009 study for Princeton University by Talya Miron-Shatz[54] yielded evidence that, among American women at least, a tendency to focus on financial matters reduces the likelihood that you'll be happy, *regardless of income*. It's not just how much money you have, it's also your *attitude* to money, hence people on six- or seven-figure incomes can still be far less happy than those earning a fraction of their salary.

We know money is recognised by the brain as a valid reward because it's needed for our survival, but unlike with food or water, there doesn't seem to be an obvious point where the brain says 'stop, that's enough'. There's technically no upper limit to how much you can earn, but to absolutely guarantee financial security in the face of every possible expense, calamity and challenge the world can cough up you'd need ludicrous sums saved up. So if someone's prone to paranoia or pessimism, they may never think they have 'enough' money, and live in constant fear of financial ruin regardless of income. Not good for happiness, clearly.

Some even argue that, because money triggers the reward pathway in the brain somewhat like a drug, then some people get addicted to money.[55] This would explain a lot of questionable behaviour you see reported, like super-wealthy tycoons still engaged in cut-throat and vicious business practices. Addiction drastically alters the brain, the regions responsible for processing sensations of reward seemingly reaching out and altering or suppressing areas like those in the frontal cortex responsible for restraint, logic and other conscious behaviour.[56] This has the effect of altering our priorities, inhibitions and motivations, meaning we become focused on the source of our addiction and prioritise it above all else.

However, addiction is further fuelled by tolerance, where the brain adapts to the source of the 'hit' so it loses its potency. This would mean you need ever-greater sums of money, and once you've exhausted all the 'normal' ways to obtain it then you're going to have to try more risky or uncertain approaches. Starting new businesses, chancy investments, stuff like that. But the financial world is rarely very forgiving, so this all increases the odds of losing everything. So perhaps it's not surprising to hear, as Kevin says, that those who pursue money above all else are at greater risk of losing everything.

With this in mind, I asked Kevin about his own motivation when starting out, if it wasn't just money for its own sake.

'I wanted security,' he replied immediately, consistent with my own investigations about how the brain does things. But this wasn't just some subconscious reflex on Kevin's part, he speaks from vivid experience.

'I set myself a target of being financially "free", so I had security and my children had security as well. Because after being homeless in 1988 I didn't want that ever happening again. That was my motivation, and I'm glad because it made me appreciate every penny.'

Indeed, this admission would certainly explain a subsequent zeal to succeed in business, and in life. The human brain is known to have numerous 'optimistic' biases,[57] meaning we regularly assume things will be fine, often for no reason. Maybe having this stripped away, at least temporarily, by directly experiencing said calamities can be an even more powerful motivator? And as Kevin says, it may well lead to greater appreciation for the little things. Maybe all the proverbs and clichés that make this point have a core of truth; maybe suffering in the past makes you happier now, because our brain has direct experience of how bad things can get in comparison?

Still, if Kevin wanted financial security, he clearly got it. So, what then? When he'd achieved his goal, what did he do after that? He did pretty much what you'd expect: ended up sitting on a beach in Barbados, drinking mojitos. Only thing was, as he informed me, by the end of the second week he was bored senseless, so he promptly returned home, got into charity work, back into his businesses, and eventually realised he *liked* his work.

Surprisingly, Kevin's experiences are consistent with what we know. Studies have described the relationship between income and happiness (or life satisfaction, or wellbeing) as curvilinear.[58] This means that happiness increases as our money increases, but only to begin with. After a point, this relationship lessens, and the same increase in money no longer produces as much happiness. An impoverished person may be overjoyed to receive a thousand pounds, a millionaire may not even notice it. One argument is that the point where this relationship changes is when all our physiological needs, our survival needs, are taken care of. If you've got ten million in the bank and no mortgage, the odds of you starving to death are negligible, and even the most pessimistic person surely recognises this. We still have the psychological needs though – for control, for competence, for approval – and there's also the need to simply stay active and engaged as

Kevin mentioned, but these aren't easy to satisfy solely with money.

It's not just about boredom though; there's a darker side to this, as Kevin observed.

'I'm from a farming background, and you see this happen a lot. People work all their lives on the farm, then they retire and live in the village, and after five or six years, they die. It's the people who stay active, stay involved, they last a lot longer. It's one of the reasons I can't see myself ever retiring.'

It's true that retirement can often strip people of their purpose, their motivation, but the immediate health consequences are surprisingly direct. Our work, whatever you think of it, offers stimulation, even if it's mind-numbingly repetitive. It also (hopefully) provides the other things discussed in this chapter, to a lesser or greater degree. But when you retire, all this suddenly stops. Sudden major life changes like this stress out our brain, which doesn't like uncertainty, and therefore make us unhappy. Also consider that we spend so long working, it becomes a big factor in our identity, especially if it's a job we like. You can spot this in the language used; contrast 'I work in admin' or 'I work in retail' with 'I am a doctor' or 'I am a pilot'. Leaving a job you like, even voluntarily, can be like losing your home, in that you lose an important part of your sense of self. It's no wonder retirement ranks very high on the Holmes and Rahe stress scale, even higher than pregnancy!*[59]

Granted, many people don't like working, so retirement has an obvious appeal. But maybe it would be better to describe retirement as stopping *earning*, rather than stopping

---

* The timings determined by nature mean it's very unlikely to retire while pregnant, but Lord alone knows how stressful THAT would be.

working? Because resorting to complete inactivity is very bad for the brain, reducing its durability and overall health. And while it's still poorly understood, the fact that things like the placebo effect even exist reveals that our mental and physical health are fundamentally linked, sometimes to a fatal degree, as Kevin noticed.

I was also intrigued by Kevin's own studies, where he travelled the world interviewing hugely successful types, to gain insight into how they achieved what they have, and wondered if there were any particularly important findings he could share. There were:

'I met all these hugely wealthy people, but I always asked them the same questions. And the first question was always, "Are entrepreneurs born or can they be made?"' The classic nature vs nurture debate, in other words. 'Almost 100 per cent of the answers were that, given the right environment, anyone can succeed.' But whether the answers given to this question were entirely consistent with reality are unclear, as Kevin himself observed.

'If you give people the right environment, they can excel, and this was proven to me many times, and it was quite profound. But it also became clear that you've got to have the spark inside you first.' This may sound inconsistent, but it isn't really. Countless people must share the same environment as a successful person, but only one of them goes on to achieve great things, so there must be something unique about them. And surely that's genetic, a 'nature' thing? But Kevin wasn't done yet.

'That spark is created through a number of reasons, but normally through pain somewhere else, and the will to succeed that often comes from having experienced that pain.'

No doubt Kevin's own experiences with homelessness loom large in this conclusion, but it would be hard to dispute it. The 'spark' needed for success may well be due to some genetic or other inherent factor – for example, Bill Gates, one of Kevin's interviewees and, at time of writing, richest man in the world, seems to have had a relatively comfortable upbringing by most people's standards, but started demonstrating an alarming competitiveness at the age of eleven, suggesting some underlying quirk of genetics or development that drove him. But said spark may also be provided by our environment, by enduring trauma and other unpleasantness, meaning our brains have very clear memory of going through that and are instilled with a deep and enduring motivation to avoid it again at all costs. It seems that one of the things that makes our brains more motivated to succeed is a direct and visceral experience of failure and despair. Perhaps we're back to the good old overactive threat-detection mechanism again?

This isn't automatically a good thing of course; maybe those more driven for success have less qualms about bulldozing anyone who presents an obstacle? But that it's a powerful motivator is beyond doubt.

This presented an interesting outcome, though: that the drive to work hard and succeed, to achieve all the things we think will make us happy, is greatly enhanced by being decidedly *un*happy. Could that be right?

At the start of this chapter, I mentioned that I somehow remained happy despite my investigative failures. Later, I revealed that I'd previously spent two grim years employed as a cadaver embalmer. Now I couldn't help but wonder; were these two things linked? Would I have been so sanguine about my faltering progress if I didn't know first-hand what it was like to be truly miserable at work? Admittedly, genuine trauma is often extremely debilitating, but as far as the brain is concerned, negative experiences *are* indeed beneficial, for our mental health, wellbeing and, of course, happiness. A greater wealth of experiences for the brain to utilise and refer to over the subsequent course of our lives can significantly boost our ambition and motivation – which, as we've seen, are closely tied in to happiness. Experiencing a wide range of possible emotions, good and bad, means we also gain greater emotional competence,[60] allowing us to react and respond appropriately. This should also lead to greater happiness, for various reasons.[61]

One obvious conclusion from all this is that, thanks to the murky workings of the brain, the factors that determine whether someone will be happy at work are fantastically complicated and variable. If this is the case though, surely the constant, often borderline-fanatical corporate attempts to essentially *force* their employees to be happy are doomed to failure? Indeed, the available data (plus talking to any employee on the receiving end of these efforts) suggests this is the case, with studies reporting that just 30 per cent of employees are 'engaged' with their work in some way, let alone happy about it.[62]

Luckily, this could in fact be a blessing in disguise. In an illuminating piece for the *Harvard Business Review* in 2015,[63] André Spicer and Carl Cederström laid out many ways in which perpetually happy employees, while maybe individually more productive, could actually be *bad* for business and the workplace. For instance, happy people aren't that good in negotiations; they often capitulate more readily, to avoid negative interactions. Angry people tend to do better here.[64]

Being constantly happy at work also means your outside life suffers in comparison, so your home life and familial relationships can become strained, cancelling out the benefits. If you're happy in your job you're still at the mercy of economic factors, and happy employees are more devastated by a job loss than others. Other concerns seem to overlap; continuously happy workers often require constant praise and positive feedback and get upset when they don't get it, and they can be lonelier and more selfish because they're more focused on staying happy than engaging with others. None of these are good for business.

Once I got home from Kevin Green's office, I tried to sit down and incorporate everything he'd told me with what I already know, to summarise exactly how and why working affects the brain to make us happy, or unhappy.

Our brains recognise that we *need* to work to survive, and working can have numerous health benefits, mental and physical. But, our brains are also evolved to avoid expending effort for no reason, so make us reluctant and averse to hard work if we don't get any obvious benefit from it. Thankfully, jobs generally mean we get paid money, and the brain recognises money as a valid reward, because we need it to survive, and the more we have the 'safer' we are, so getting more money often makes us happier. It's more complex than that though, as the human brain is sophisticated enough to want to do more than just survive, meaning we have other needs, psychological rather than biological. These include a need for autonomy, to be in control, to feel competent, to be appreciated and approved of, to achieve our long-term goals. A job that can provide any or all of those things is far more likely to make someone happier at work. An absence or active removal of these things, especially if still demanding they do their jobs to an unreasonable standard, will make a workforce deeply unhappy.

In 2015, the *Guardian* looked at multiple surveys which asked what jobs make people happiest,[65] and their findings are largely consistent with my conclusions. Apparently the happiest job was being an engineer, a well-paid role which includes autonomy, competence, and a very tangible way to observe the outcomes of efforts. Other jobs mentioned include doctors, nurses, teachers and, perhaps surprisingly, gardeners.* What didn't feature at all were jobs like call-centre operator, shop assistant, fast-food worker, things like that. These jobs are often highly demanding with few rewards and poor pay. These are often positions used by big, wealthy businesses though, so maybe the idea that happy employees aren't good for business is true after all, and those companies that insist on them are wasting copious time and money?

Why *would* happiness be so detrimental, though? Well, recall how, in the previous chapter, we saw how we need to engage with others, but because that takes cognitive effort we also need privacy and space to give our brains a break, let them 'recharge'. It's a similar story with happiness. While it may take many forms, the 'productive' sort of happiness, where someone is upbeat, cheerful and motivated, is bound to prove exhausting for the brain if we keep it up too long.[66] This metabolic price of happiness could mean that the brain ends up prioritising it over other important things, like being

---

* Or perhaps it isn't surprising. It also provides autonomy, competence, visible results of efforts, and don't forget the psychological benefit of green spaces mentioned in the previous chapter.

generous and considerate, which negatively affects us overall.

There's an important point to this; we need to work, but we technically don't need to be happy. We're meant to be happy when good things happen for us, or we're doing things we enjoy. But by insisting on constant happiness, as many workplaces and even much of modern society seems to, we're throwing things out of balance, simultaneously cutting our brains off from a more diverse range of emotional experiences, and overtaxing it.[67] It turns out the work–life balance may be more valid than I previously thought, but the key word isn't 'work', it's the 'balance', which too many overlook, much to their detriment.

What it may all boil down to is, while it may indeed be possible for your work to make you happy, the reason it's so difficult is that, from the brain's perspective, happiness often *is* work!

I admit, I had to have a little sit down when I realised that.

# 4

# Happiness Is Other People

You may recall from the previous chapter that my interest in the role of work and jobs in our happiness was inspired by a visit to a cutting-edge art gallery in the ancient Italian city of Bologna. However, to show that my life isn't all exotic locations and profundity, everything covered in this chapter began with a sandwich.

As I was driving home from my meeting with Kevin Green I realised it was lunchtime, so I pulled into a nearby retail park and went to a well-known sandwich franchise. However, whilst queuing, I realised that I actually wasn't far from my mother's house. With prior planning, I could have gone there for lunch instead. Ah, well. However, because I was lost in this train of thought, when the woman serving me said 'Enjoy your food', I replied with: 'Thanks, Mum'. After the brief-but-excruciating pause that greeted this, I ended up running out the door, mortified. The embarrassment! What if she thought I was being sarcastic? Or derogatory? Or experiencing some sort of Freudian mental breakdown? What if everyone in the shop was right now laughing at my idiocy?

Eventually, the scientist part of my brain kicked in, and said *so what*? Worst-case scenario, some strangers experienced fleeting amusement at my weird-but-harmless faux pas. In practical terms, none of it mattered. Except . . . it *did* matter! Minutes earlier I was happy; now I was cringing myself inside out in a rain-soaked car park, holding an increasingly soggy sandwich, all because of one minor exchange with a stranger. This objectively inconsequential incident had had a rapid and substantial impact on my happiness.

But maybe I should have expected this? In everything I've covered so far about what makes us happy, there's been a persistent element I've alluded to but not yet tackled outright: the impact of other people. Why do we aspire to a nice home? Vast wealth? To be a sporting champion? Many reasons, but underlying all of them is a desire for the approval, admiration and respect of fellow humans. On the other hand, other people can be a source of considerable unhappiness: toxic co-workers, fraught domestic situations, estranged family, sandwich artists maliciously tricking you into saying something embarrassing, all these things that can make you very unhappy indeed.

Clearly, when it comes to being happy, our brains place considerable value on positive interactions with, and the approval of, our fellow humans. Exactly how and why this is the case is what we will consider next. Assuming nobody minds, of course.

## Evolution, a friend to intelligence

Neurologically, my reaction to my public blunder was rather telling. I logically figured out there were no lasting consequences from it, but by that point I'd experienced deep and visceral embarrassment regardless. Clearly, our brain responds to social interactions in ways that are separate from, and faster than, conscious thought. The intelligent, logical parts of our brain only *limit* the mood-rattling effects of awkward public mishaps, not *prevent* them. It's a bit like a wise old man explaining to the exasperated firemen how his curious grandson accidentally set off the alarm.

Such setups are usually reserved for things that are deemed important to our survival. For instance, human attention is controlled by both 'top-down' and 'bottom-up' processes. Top-down is where we consciously direct our attention towards things we want to focus on.[1] When you scrutinise the different parts of a *Where's Wally?* picture, trying to find the elusive bobble-hatted character, you're using top-down attention. Bottom-up is where the more reflexive, instinctive elements of the brain detect something 'biologically significant' in our perception (a possible threat, a potential reward, an attractive mate, etc.) and immediately shift our attention towards it.[2] If you're sitting alone watching TV and the aforementioned Wally jumps out at you from a cupboard, the bottom-up attention system will be shifted towards him, whether you want it to or not.

Unfortunately, for the brain, complexity means delay, like ordering an elaborate cocktail at a busy bar. Hence the top-down system is often slower to respond. When you're alone at home and a book falls off a shelf, the sensitive threat-detection systems that guide bottom-up attention immediately say 'UNEXPECTED NOISE! POSSIBLE MURDERER!', meaning your heart is hammering before your conscious, analytical processes can work out what actually caused the noise. As my own experience showed, it seems to be the same when we mess up social interactions. Also, consider that when we're embarrassed, our face invariably goes bright red. Do we *choose* to do that? Do we stop and consciously think, 'This faux pas could be improved by me resembling a mortified tomato'? Of course not. It just shows there's a clear subconscious, involuntary element to our social interactions.

Similarly, if we eat spoiled or unpleasant food, the feeling of disgust is instant, powerful, involuntary, and very *enduring*; you get poisoned by a tuna sandwich, you can tell yourself it was an unfortunate one-off all you like, it'll still be a long time before you eat tuna again. The brain has dedicated areas for processing the disgust response, like the insula,[3] because eating spoiled food is a genuine hazard, and our brains have evolved mechanisms to prevent it.

Now, think of an embarrassing event from your life. An alcohol-induced insulting speech at a wedding, a humiliating 'wardrobe malfunction' at a school dance, saying something ridiculous about Robocop in a workplace workshop; when you make a spectacular fool of yourself, as we all do at some point, do you ever really 'get over it'? We tell ourselves that nobody cares or remembers these incidents, but they still cause a deep and enduring sense of shame whenever we think of them, like the residual nausea if we even think about a food that poisoned us.

Why? The spoiled food and poisoning thing makes sense, but why would we be so helpless in the face of potential disapproval from others? It's not as if other people liking us is a matter of life and death. Interestingly, and somewhat ironically in this context, the answer lies in how our brains ended up being capable of such rational thought and analysis in the first place.

Huge brains and vast intelligence are no inevitable consequence of evolution. Big, smart brains are an incredible drain on resources, and anything that saps energy for no reason is a no-no for natural selection. You could feasibly install a huge supercomputer in your car, making it the smartest car in the world. But why would you? It'll just be a massive drain on fuel and battery, and you'd be regularly overtaken by simpler but more efficient cars. And so it is with brains and evolution; organisms are invariably only as intelligent as they *need* to be.

So why did we humans end up with brains considerably bigger than our body size would suggest,* using around 20 per cent of the body's available energy just to stay alive?[4] Weirdly, evidence suggests that in the last three million years the human brain expanded in size by around 250 per cent, much of which was concentrated in the intelligence-producing cerebral cortex, and most of it happening in the last one-and-a-half million years! Whatever made us so smart also happened relatively recently. On evolutionary timescales, it's like Peter Parker being bitten by a radioactive spider and waking up the next morning with super powers. So, why did natural selection favour big brains in humans? What, if you will, was mankind's radioactive spider?

Some argue that it was because intelligence was a sexually attractive trait,[5] as it suggests health, good genes and disease resistance, so intelligent humans got to mate more, spreading their intelligent genes, raising intelligent children, and so on, in a self-perpetuating cycle. However, if this were all there is to it, we scientists would be the sexiest people in the world today, and clearly that's not the case.** Instead, many theories such as the social brain hypothesis[6] and the ecological dominance–social competition model[7] argue that the biggest factor

---

* It's a persistent myth that a bigger brain automatically means greater intelligence, but a higher-than-average brain-to-body ratio (i.e. having a comparatively larger brain in relation to body size than species averages) is a much better indicator.
** With notable exceptions, like myself. Most days I can barely move for groupies. It makes the school run very tricky.

was our sociability; our desire to form relationships and gain the approval of our communities.

Think of what being part of even a basic human tribe involves. You have to know who is who, observe rules and social norms to maintain peace and be accepted within the community, coordinate your actions with others in activities like hunting, defence, foraging, etc. You need to look after the vulnerable, or repay those who provide for you. You need to form alliances and relationships, and resolve disputes when this fails. You essentially must maintain an up-to-date network of links, alliances and histories, as well as reliable real-time simulations of many other humans, by far the most complex things in any environment, *all inside your own head*! This requires considerable reserves of brain power. Luckily, we humans have them.

Many other animals also form social groups, and indeed, the evidence suggests that the more social an animal is, the more intelligent it tends to be.[8] However, this isn't an absolute. Solitary tigers, for example, have a higher brain-to-body ratio, suggestive of greater intelligence, than social lions. Also, 'simpler' creatures like rats, mice, and even wasps,[9] form recognisable social groups. Being part of a social group may be easier if you're intelligent, but it's not *essential*.

However, another factor that links social interactions to intelligence is mating strategy. Most animals are promiscuous, which doesn't require much intelligence; it's all about recognition ('Look, a sexually attractive female!'), fertility cycle ('And she appears to be in oestrus!') and access ('I wonder if that hulking brute standing over her will let me mate with her?'). It is a process more reliant on pheromones and opportunity than thoughtfulness.

By contrast, pair-bonding – nature's equivalent of monogamy[10] – requires serious thought, as anyone who's forgotten an anniversary will know. It means incorporating a whole other individual's needs, situation and behaviour into your own thinking. It's essentially a complex social group, but with two members. Accordingly, many mammals and birds show a correlation between increased brain size and intelligence, and a tendency to form lifetime pair bonds.[11] Basically, monogamous animals are smarter, because they *need* to be.

However, species that rely on pair-bonding often have dedicated neurological systems that encourage and reward such behaviour, so greater intelligence isn't 100 per cent essential. Oxytocin and vasopressin (another of the brain's so-called 'happiness chemicals') are a big part of the brain's ability to 'bond' to one specific partner. Sensory cues linked to the individual's partner (their face, body shape, scent, etc.) trigger these chemicals, which in turn trigger dedicated receptors in the mesolimbic reward pathway via dopamine and other neuropeptides. It's complex and multi-layered, but it basically means individuals experience pleasure when engaging with their partner, and conditioning processes mean that it eventually associates the mere sight of them with reward, with pleasure, with . . . happiness?

Admittedly, this system as described is derived from studies in mice, although evidence suggests more sophisticated mammals, like primates, are beholden to similar neurological processes.[12] But still, if your species is already reasonably smart and lives longer than the single year or so mice can expect, greater intelligence will be needed to maintain pair-bonds. And in numerous species, there's evidence linking pair-bonding to increased brain size.

Indeed, many argue that humans becoming monogamous was a key step in our intellectual development, and at least one theory suggests that in humans (and other primates) the neurological mechanisms supporting pair-bonding were somehow 'detached' from the mating process,[13] meaning we could form long-lasting, emotionally-rewarding bonds with *multiple* individuals, not just reproductive partners. Basically, we developed the concept of 'friends'. And if forming lifetime connections with just one individual requires bigger brains, what if you do it with several? Dozens? *Hundreds?* Your brain needs to be exponentially more powerful again. Hence, in primates specifically, size of the typical social group is strongly linked to brain size and intelligence.[14]

Yet as smart as our primate cousins can be, we humans are considerably smarter again. The predominant theory about why is the aforementioned ecological dominance–social competition model, which argues that human social groups got so successful that the usual ecological pressures driving evolution no longer applied; if you're part of a human com-munity, you're protected from things like predators and have ready access to food, safety, mates, etc. So, succeeding in the *environment* was less important than succeeding in the *community*. Survival of the fittest now means survival of the most likeable, friendly individuals, able to benefit the group with ideas and innovations like tools and agriculture. These individuals were the ones who succeeded, got to spread their genes. But all those things require greater intelligence. Sev-eral hundred thousand years later, here we are.

The point is, thanks to how we evolved, sociability is deeply embedded in our thinking, our consciousness, *our DNA!*[15] Even comparisons between humans and chimpanzees (our closest evolutionary cousins) show this: tests reveal chimps are better at visual, sensory processing than we are, whereas we're far better and more inclined towards social processing.[16] Essentially, if you give a chimp a banana, it'll focus on the banana. 'A banana. I like bananas. I'll eat that.' If you give a human a banana, they'll focus on *you*. 'Why's this person giving me a banana? What do they want? Are we "banana buddies" now?' and so on.

This is what happens when a species evolves according to social rather than environmental pressures. If your survival depends on your community, your group, the more social you are, the greater your chances of acceptance and survival. Being shunned or rejected by the group, that's no little thing; in the hostile world we evolved in, it's tantamount to a death sentence.

That's why our supposedly logical brains treat being accepted by others as a matter of life and death; because as far as they're aware, it is!

## Keeping in touch

You may think, sure, social interaction got us to this point, but we've moved beyond that now. Our interactions with other humans might have made us more intelligent, but we don't exactly 'need' other people to make us happy nowadays, any more than we need stone axes to butcher a gazelle for dinner. And you'd have a point: our increasingly sophisticated technological world means we can now work, eat, sleep and play in ways that involve little to no direct human contact. So, what difference do our social interactions make to our happiness?

Quite a lot, as it happens. Remember, our brains have undergone rapid, intense evolution over millions of years, seemingly driven by a pressure to make us as friendly and interactive with our fellow *Homo sapiens* as possible. This had profound and lasting effects on our brains, which haven't just gone away because we've invented Netflix and pizza delivery. Our brains still have many systems, circuits, processes and mechanisms, both conscious and unconscious, dedicated to facilitating and encouraging connections and exchanges with our fellow persons. So yes, we *can* live our lives and even experience happiness in the absence of other people, effectively bypassing all these neurological systems, but we could also get about purely by hopping, bypassing one of our legs; it's possible, but it's much easier and less harmful *not* to do that. The point is, other people aren't just another part of the environment, like trees, buildings and bus stops; something we interact with via our senses that our brains react to as and how the context requires. They're a major factor in how our brains work.

For instance, most social species are gregarious, actively seeking out others with whom to interact. It makes sense; social bonds may be important for your survival, but they don't just 'happen'. They take time and effort to forge and maintain, as anyone who's had a BFF* at school they now rarely speak to will realise. To that end, our brains have evolved to encourage active friendliness. For instance, we've seen how oxytocin encourages and rewards social interaction. On top of this, a 2014 study by Lisa A. Gunaydin and colleagues[17] provided evidence for a specialised circuit linking the ventral tegmental area (just by the brainstem) and nucleus accumbens to the

---

* Best Friends Forever, if you didn't know. An extremely optimistic projection in any context, given how humans aren't immortal.

lower frontal regions of the brain that encode and predict social behaviour. Increasing or decreasing activity in this specific circuit (in mice, admittedly) caused a corresponding increase or decrease in social interaction behaviour. And if these brain areas sound familiar, they should; they and the neural connections between them form the mesolimbic reward pathway.[18] The mere act of interacting with another person can be enjoyable, and with good reason; the mechanism that guides our desire for social interaction is embedded right in the part of the brain responsible for the experience of pleasure. It's like a party invitation, wrapped in bundles of cash and hand-delivered by all your teenage crushes; you can decline, but it's not easy. No wonder we're so keen to stay in touch with others.

I use the term 'in touch' purposefully, because experiencing happiness from social interactions began with physical contact. Specifically, grooming. Most animals groom themselves, to remove dirt and parasites from their skin/fur/scales/feathers, etc. Some spend hours doing it, like self-indulgent cats. It's good for hygiene and health, so evolution has made it *feel* good. Literally. Touch is experienced via nerves in our skin that respond to changes in pressure (and more[19]), which send corresponding signals to our brain. Some of the neurons facilitating this are called c-fibres, which are smaller and slower at conducting signals than many other neurons.[20] They transmit sensations like dull, aching pain, but also convey pleasurable touch. While all touch sensations are processed by the brain's somatosensory cortex, c-fibres also convey *pleasant* touch to the insular cortex, a region associated with pleasurable sensation and reward-seeking behaviour, particularly in drug abuse.[21] And, thanks to evolution, one such pleasant form of touch is grooming.[22] Ever wondered why humans pick scabs?

Or their noses? Such behaviour is pointless, unless maybe the old brain circuits that reward the act of removing debris or superfluous matter from our physical form are still in there somewhere? This may well also explain why some people bite their nails when stressed.

However, most social species form and maintain bonds by *social* grooming. It's more enjoyable if someone else does it. This may be because the sensation of being groomed has no corresponding motor cortex activity in the brain – like how you can't tickle yourself, because your own brain 'knows' tickling is coming.[23] If someone else does it, that's way more unpredictable and intense.

Social grooming has similar traits. Indeed, it even releases endorphins, producing feelings of relaxation, pleasure and happiness.[24] Give a social animal opiates, they lose all interest in grooming because they're already experiencing the 'buzz' they get from it, but inject something that blocks the action of endorphins, they desperately crave it.[25] Seems you can be genuinely hooked on social grooming. Also, social-grooming chimps show greater levels of oxytocin, crucial for feeling the rewards of interpersonal bonds, after being groomed by individuals they're already linked to, like bond mates, kin and members of their own group. Ergo, social grooming forms bonds, but also cements existing ones. It's got to the point where many animals, like baboons, spend much of their waking lives social grooming, far beyond what's needed to maintain hygiene. Apparently, the species that grooms together, stays together.

However, this presented a problem for humans, as maintaining bonds via social grooming takes time and effort; the more individuals in your group, the more grooming time needed. And human groups kept getting bigger. What to do?

One theory is that humans adapted their existing verbal\*
communication and language skills to effectively replace
social grooming. Basically, rather than spend hours picking
the ticks out of someone's fur, we could say variations of 'I like
you' instead. Our brains seemingly respond to compliments
and praise like they would to social grooming,[26] except it's a
lot quicker and easier, and can happen at a distance.

If our language and communication abilities have been
co-opted to facilitate social interactions and relationships, is it
any wonder we spend so much time gossiping with friends in
pubs and coffee shops? Some suggest that gossip is the very
reason we developed complex language in the first place.[27]
The use of language to reinforce social bonds, coupled with
the brain's inclination to gather useful information, means
discussions that reveal details about others in your group/
community/society are especially rewarding, as sales of tabloids and those omnipresent magazines reveal. Indeed, like
social grooming, the amount of time we spend nattering in
the coffee shop or pub is far beyond what's needed to convey information. At least gossip is verbal, and we're not still
picking parasites off each other's bodies. That'd make the
Starbucks visits very awkward, if nothing else.

That's not to say physical touch isn't still important for
human interaction. Hugging, handshakes, foot rubs, pats
on the back; humans have a variety of ways to cement positive interactions via touch. It can be surprisingly potent; one
study suggests waiting staff receive greater tips if they casually touch the customer in some way.\*\*[28] But still, for human

---

\* And visual; many argue that spoken language developed from physical gestures,
and indeed sign language seems to activate the same brain regions as spoken ones.
\*\* Meaning a touch on the shoulder or brushing hands when taking orders, not

social interaction, touch is supplementary to language, not the 'main event', so to speak.

Some may find all this unsettling. We often like to think we're strong-minded, independent individuals, so the idea that simple communication can be so potent that our brains, feelings and moods can be so easily affected by other people is somewhat disturbing. Well, strap in, you ain't seen nothing yet.

## The lives of others

EEG studies have revealed networks of neurons (dubbed the 'Phi complex' in the right centro-parietal cortex) that display patterns of synchronised activity when two people are interacting. The gist seems to be that these brain regions essentially form 'hubs' in an 'interindividual brainweb' as I've seen it described in a published scientific study[29] – and not, despite how it sounds, a nineties cyberpunk novel.

The irony of something describing fluid communication between two humans sounding like jargon-riddled nonsense is not lost on me, so what does it mean in plain English? Well, the phi complex is a part of the brain that specialises in processing personal interactions, in real-time. It's activated when two people interact, whatever form this interaction takes. But the interaction itself is one *thing* being created by two brains, so part of both brains will effectively be 'synchronised': they're both processing the exact same information. If you show two different brains the colour red, both will show very similar activity in the retina right through to the visual cortex.[30] Think

---

aggressive groping or bear hugs. That probably wouldn't make someone want to tip you.

of it like two modern video game consoles, playing a game together online. The interaction is the game, the flow of sensory information is the online connection, the consoles are the brains, and the phi complex is the representation of the game in each console.* The point is that when two people interact, their brains effectively 'synchronise'. Which is cool. Or alarming. Your mileage may vary.

This process is believed to be supported by mirror neurons. In the 1980s, neuroscientist Giacomo Rizzolatti's research team were studying activity in the motor cortex of monkeys when they discovered that the neurons which fired when a monkey reached for or bit a peanut also fired when the monkeys *saw someone else doing those things*.[31] Admittedly, individual mirror neurons haven't been located in humans yet (it was mostly luck that they were found in monkeys) but there are what appear to be mirror areas, which perform the function and show the sort of activity you'd expect if there were mirror neurons in them.

While mirror neurons in other creatures allow them to mimic and learn from others, humans have seemingly taken things further. Ever found yourself wincing in sympathy when someone describes a gruesome injury they've suffered? Did you inwardly cringe on my behalf when reading about my sandwich incident? Do you feel angry when hearing about an injustice someone has endured? Why? None of these have any bearing on you, yet you still react emotionally as if you were personally affected. And you're not just feigning it out of politeness; studies have revealed that people observing

* Technically minded readers are probably screaming right now about servers and processors and so forth. In my defence, I'm just a neuroscientist; if it doesn't splat when you drop it, I'm not much use to you.

someone smelling something unpleasant will show activity in the areas of their brain that process disgust,[32] and that when our brains read facial expressions the emotion the expression reveals prompts neurological activity in the areas that process that emotion in ourselves.[33] This is empathy, the ability to understand and share the feelings of others.

The automatic, unthinking process where we share other people's emotional experiences is called *affective*, or emotional, empathy. But there's also *conscious*, or cognitive, empathy, aka theory of mind,[34] which is the capacity to consciously understand someone else's mental state, to realise they have their own sophisticated inner life, different to our own. No other species seems able to do this (within reason[35]), whereas human children pick it up quickly.*[36]

While they usually overlap, these conscious and unconscious empathy processes can be distinct. You're telling someone about why your job is awful; they gasp, sigh, shake their head at all the right points. They clearly empathise with your plight. And then they say, 'Well, why not just quit?', as if you're so dumb you haven't considered that and ruled it out. This person has good affective empathy, but poor conscious empathy, and their response is more annoying than helpful. Similarly, someone may listen to your story with what seems like utter disinterest, then offer a perfect solution. Their conscious empathy is sound, their affective empathy, not so much.

The implications of this are many and profound, but one obvious one is that we can essentially *share* happiness. It can spread. Many things can make us happy, like fine dining,

---

* You may have heard of this in the context of autism. Some argue that people with autism have some form of deficit when it comes to mirror neurons and ability to achieve theory of mind.

exploring exotic locations, creating artworks, working on your home, going to the theatre or cinema, playing sports, and so on. It's possible to do these things alone, but people rarely do; having someone else along to share in the experience is a big part, sometimes even the main part, of what makes it so enjoyable. And part of this may stem from the fact that our brains allow us to 'experience' the happiness of others, as well as our own. So, we do something we enjoy, which makes us happy, and if we're with someone who also enjoys it we empathise with them, which makes us happier, plus our brain rewards us for our social interactions, which makes us happier again, and so on.

The overall point is, a large part of our brain is dedicated to encouraging and facilitating social interactions. This would point towards social interaction being a basic requirement of a healthy brain, and not just a pleasant bonus. Logically then, a lack of social interaction would be genuinely unhealthy. And this seems to be the case. Animal studies have shown individuals that don't experience social interaction readily develop psychological problems and disturbances.[37] Not only that, studies in monkeys have shown that the brains of individuals raised in isolation are noticeably different to those raised in company.[38] Worryingly, a lack of social interaction causes clear, detrimental cellular and even chemical changes in the regions responsible for processing reward and pleasure. This would suggest that not only does social interaction make you happy, a lack of it can make it harder to even experience happiness! It's no wonder psychologists consider solitary confinement to be a form of torture.[39]

So, logically, if you want to be happy, just interact with as many people as possible, as often as possible. As long as there isn't anything more to it than that, that should work.

Unfortunately, there is. So it won't.

# If everyone jumped off a cliff

As a child, I once asked my mother if I could go and play by the river near where we lived. She said no, because it was 'too dangerous'. I countered by saying 'everyone else plays there', which, in fairness, they did. Mother then fell back on the classic parental response of: 'If everyone jumped off a cliff, would you do that?'

My reply was, 'Well, given that the human brain has evolved an intrinsic need to be liked and accepted by others which can and often does overrule our rational decision-making abilities, even if it has obviously negative consequences for our own wellbeing and even survival, if I was confronted by a scenario where all of my friends, none of whom have ever jumped off a cliff before, all suddenly and simultaneously decided to do exactly that, then I can't promise I wouldn't observe this and assume there was a valid reason for their behaviour, prompting me to follow suit. To summarise, given how the brain works, yes I probably would jump off the cliff.'

Well, that's what my reply would have been, if I'd had more time to think about it. Say, twenty-five years more. But regardless, it's the truth. The cliff-based quandary posed by parents the world over is far from the rhetorical question they think it is.

Thanks to the way our brains work, social interactions are something we seek out and enjoy, but then, we seek out and enjoy food too. But not constantly, and it doesn't automatically follow that because we *can* enjoy food we will enjoy *all* food. And that's true for social interactions. For instance, someone tried to mug me once. He failed, mostly because he was half my age, and size, but technically, this counts as a social interaction. I didn't enjoy it though, and I'm guessing neither did he. It needs to be a *positive* social interaction. What makes a social interaction positive? Well, many things, but most of them boil down to making the other person *like* us in some way. A shared joke, some interesting gossip, a productive meeting, a pleasant commercial transaction, or even a display of compassion during hard times, like attempting to console or help someone following the death of a loved one; whatever the conscious motivation behind them, these all increase the likelihood of the other person in the interaction thinking well, or better, of us. Because the brain wants, *needs* us to be liked, or at least accepted, by others.

Look at biker gangs, goths, punks or skinheads. Be it imposing leather, black-only outfits or elaborate hairstyles, there's a specific image or aesthetic that they all tend to adopt. Often these are people who have actively denounced the expectations, standards and even laws of wider society, yet they still obey a dress code. Why? Because, for all their conscious rejection of the demands of the wider world, the human brain's need for acceptance from others runs very deep.

A lot of this seems down to activity in the striatum. Earlier I mentioned that a lack of social interaction causes deficiencies in certain areas of the brain responsible for experiencing reward and pleasure. The striatum is one such area, arguably the main one. It includes the nucleus accumbens, earlier described as a crucial part of the brain circuit that prompts social interaction and the general ability to experience pleasure. In essence, the striatum is the part that makes us feel good about social interactions, as and when relevant.

For instance, one interesting study looked at people's

behaviour in a scenario where they could donate money to charity or keep it for themselves.[40] The results show that people are far more likely to donate to charity when others are watching them, and that there's a noticeable increase in activity in the striatum when doing so. You might argue that this is more to do with avoiding condemnation from others, rather than feeling rewarded by their approval, but the experiment also revealed that the striatum showed the same type of activity if subjects take the money for themselves when nobody is around. This strongly suggests that our brain processes social approval as a reward, one at least on a par with financial gain, as all of these induce activity in the reward-processing striatum. The study even goes so far as to argue that our brain processes financial and social rewards in the same (or at least similar) ways, so we can get the same sense of pleasure and satisfaction from both. This would help explain why living to help others can make you as happy,[41] if not happier, than just pursuing money, as Kevin Green observed.

You might think this paints a cynical picture of human nature; that we're intrinsically selfish if we have to think about whether anyone's watching us before we do something altruistic. However, evidence suggests there's not much 'thinking' involved at all. Similar studies reveal that people are more generous with tips and charity if, when given the opportunity to be so, there are just *pictures* of eyes visible.[42] In one study, subjects even became more generous when confronted with three dots, arranged in the vague shape of a face.[43] The brain's fusiform face area is the part of the visual cortex dedicated to face recognition, and it's extremely sensitive – hence people seeing the face of Jesus in burnt toast. An arrangement of three dots seems sufficient to set it off, and this in turn influences our pro-social behaviour. This shows again that the effects of social interactions on our brains go way deeper than conscious thought. It also suggests that old-fashioned types complaining about the use of emojis and emoticons in modern-day communication are in the wrong, because seeing a simple face seemingly makes us nicer, more considerate people. Constant exposure to emojis could be making the whole human race happier!

It works the other way too; social rejection is potent and unpleasant. Another neurological region impaired by an absence of social interaction from a young age is the amygdala, responsible for the sensation of fear and integral to our threat-detection systems, suggesting the negative aspects of social interactions can be equally essential for development of a healthy brain.

It's no wonder negative social interactions are considered unpleasant enough to be labelled a threat; social rejection is *painful*. Literally. Just like how positive social interactions trigger the brain's underlying reward system, social rejection seems to trigger the regions responsible for processing pain. Actual pain. A study involving a simulation where subjects played a ball game and were gradually rejected by other players showed raised activity in the anterior insula and anterior cingulate cortex, cortical regions that are linked to the experience of pain.[44] For a while, it was argued that this showed social rejection causes the same sensation as pain from physical injury, but closer analyses of the data suggest the same regions are activated but in different ways,[45] like a pen being used to write a love letter, and then a ransom demand; it's the same thing performing similar, but distinct, functions. But regardless, nobody's arguing that social rejection isn't genuinely painful, in the 'psychological discomfort' sense. I've mentioned this before, but that saying about sticks and stones is totally wrong; being called names *does* hurt. Science says so.

It doesn't even need to be anything significant; some studies reveal that we instinctively dislike people who fail to make brief eye contact when passing, who 'blank' others.[46] And the pain of social rejection persists even when it makes no logical sense; the simulated ball game study showed that African Americans still experienced the hurt of rejection when they were told they were rejected by members of the Ku Klux Klan! And people still felt hurt in scenarios where they were financially rewarded for every rejection.

As a result, our brains do everything they can to avoid rejection. We've seen that our brains are capable of self-appraisal, so we could easily present an honest image of ourselves to those around us. However, that's risky, because what if they dislike the sort of person we are? Better to 'tweak' or 'exaggerate' our good points, so we come across better. And this is what the brain does, to the extent that it often counts as self-deception. There's a process our brains indulge in termed 'impression management', where we try to give the best possible impression of ourselves by influencing the perception of others. One study into the neural correlates of this process made subjects present themselves, inaccurately, in positive or negative ways, necessitating a degree of self-deception. The results recorded increased activity in the medial prefrontal cortex and left ventrolateral prefrontal cortex.[47] But the most interesting part is that raised activity was only seen when subjects had to deliberately present themselves negatively. If they had to provide positive representations of themselves, there was no change in activity. Remember, the brain is never 'off', it's always active, like the noise of the engines on an aeroplane in flight, and what fMRI scans like this one show are *changes* in activity, reductions and increases. It's not a clear-cut thing by any means, but no change in activity when subjects are made to present themselves in inaccurate but positive ways suggests that *that's what their brains were doing anyway!* It's our 'default state'.

It's no wonder really, given how much importance our brains put on the acceptance and actions of others. And if you still doubt that, consider the following scenario: you're just getting into the shower and your bathrobe falls off, leaving you naked. No problem, that's perfectly normal. Necessary even. Now swap 'getting into the shower' for 'accidentally wandered into the busy hotel lobby'. That's not so innocuous, that's almost apoc-alyptically embarrassing. My sandwich-based mother-blunder has nothing on that. But it's the same action both times, the same process, the only difference is that now other people can see it. And judge you. And find you lacking.

This is embarrassment. It's a social emotion, an emo-tion that depends on the thoughts, feelings or actions of other people, whether we experience them direct, remem-ber them, anticipate them or even imagine them. My own embarrassing anecdote at the start of this chapter was expe-rienced first-hand, and it was awful. But it still feels awful whenever I remember it, as embarrassing memories tend to. I can never go back in that store now lest they remember me, so I also anticipate embarrassment. And I still feel a vague sense of dread whenever I go into other stores belonging to that franchise, because I imagine word might have spread. One simple exchange has caused deep and lasting emotional fallout. Similar social emotions are guilt, jealousy, grief, and so on. They are only triggered in the context of other people. So important do our brains think social interactions are, they've evolved specific, dedicated emotions to regulate them! Thankfully, happiness doesn't seem to be one of these, although as we've seen, it's a lot easier to be happy with other people than without.

Perhaps inevitably given all this, the people we relate to and interact with play a big part in our sense of self, our identity. Scanning studies have revealed that when we contemplate being part of a group or think about those we identify with, we see raised activity in areas like the ventromedial prefrontal cortex, and the anterior and dorsal cingulate cortex.[48] But these areas also show raised activity when we think about our sense of self.[49] The implication is that the groups and communities we belong to are a key part of our identity. This shouldn't be surprising; we saw earlier that our possessions and homes inform our identity, so it'd be weird if the people we surround ourselves with didn't.

That those we interact with play such a big part of how we define ourselves also explains why positive interactions and approval are so rewarding, and why rejection hurts so much. That, on top of everything else the brain does to make us socially likeable, explains why we're so susceptible to the actions, behaviour and even the moods of those around us. It's a complex and variable process, of course, but it's fairly common to everyone. We feel angry when part of a charged crowd, we assume others know more than us in ambiguous situations,[50] and often follow their lead, even when it goes against our better judgement, or our wellbeing. Unenthusiastic conscripts still march into battle with those around them, the media is advised to be careful when reporting suicides for risk of copycats, and if everyone around us suddenly jumped off a cliff, odds are we would too.

Because we want to be liked, we want the approval of others, and so we do our best to fit in. Because it's a big part of who we are. Because it makes us happy.

## What price fame?

When my first book came out, I was invited to do a promotional talk at the Aye Writes festival, in Glasgow central library. It was a popular event, eventually being moved to the main hall, to accommodate everyone who wanted tickets. I had addressed some of the stuff about humans and their love of social interaction and approval in said book, and made this point during the Q and A, saying something like 'maybe that's why people like being famous?' The chair then wryly replied 'But Dean, you should know; *you are famous.*'

Was I? *Am* I? I certainly didn't feel famous, still don't, given my daily life is largely writing, working and childcare. But then I was sitting in front of a crowd of hundreds who all wanted to hear about and buy a book I wrote, which enough people have liked since to warrant writing this second one you're reading now. I really try to shy away from anything boastful or needlessly self-aggrandising, but I'd be lying if I said none of that affected me. Earning the approval and interest of so many people certainly did make me happy, and does now. But, famous? I would still dispute that.

But thinking back to this event, something struck me: what is fame, if not the approval of many others, something that makes us happy, on a much larger scale than the average person can expect? And here's the other thing: earlier we saw that the brain processes social approval and financial gain in roughly the same way. We often talk about people being 'rich and famous', but as far as the brain's reward system is concerned, there's not that much difference between them. This helps explain those stories you hear about well-known pop bands who earn surprisingly little because they're put together by merciless music industry executives who keep all the profits. You'd think nobody would agree to such an arrangement, but maybe fame is sufficient reward by itself?

In fact, maybe fame is even *more* rewarding than riches? We've seen how our brains perceive money as a viable reward, like food or shelter, because it's important for our survival, but approval from others seems to operate on many other levels of cognition, and is seemingly important for the health of the brain. Saying that, we saw in the previous chapter how money makes us happy *up to a point*, after which its potency lessens and we find pleasure in other things. If the brain processes them in similar ways, would this be true of fame as well? Does being slightly famous make you happy, like it did with me, while being hugely famous doesn't? If it does, this could explain a lot about how the brain processes social approval.

However, to investigate if incredible fame makes you unhappy, I'd obviously have to speak to someone incredibly famous. The thing about incredibly famous people, though, is that you can't just wander up to them in the street and ask for a favour. Their fame, the very thing that I needed to ask them about, means they're kept away from people like me. What to do?

Well, to cut a long story short, after a prolonged exchange of messages between myself and well-known types on the

Welsh entertainment scene, I ended up in a quiet bar in the Cardiff Millennium Centre, sitting at a table opposite Charlotte Church, who was eating a bowl of cawl (traditional Welsh stew) while holding a copy of my previous book, *The Idiot Brain*, that I'd handed over upon her arrival like the world's most bloated business card.

In case you don't know Charlotte Church (which would, admittedly, undermine her presence in this context some-what), at the age of twelve she achieved international fame with her debut album *Voice of an Angel*. It seemed a twelve-year-old soprano was exactly what the world wanted in 1998, and the album sold millions of copies, resulting in her performing for presidents and in major movies, singing alongside megastars, hosting TV shows, and more. And now she was sat across from me in a bar. Eating stew. It throws your mental processing for a loop, I'll say that much.

Essentially, for more than half her life (she's only thirty-one at the time of writing), Church has been very famous. Her life has not been what anyone would recognise as 'normal', but has it been happy? That's what I wanted to find out. And she was generous enough to be willing to help me. So, I started with what I felt was the obvious question; did she *want* to be famous at age twelve? Did she have any concept of what it even meant?

'No, not at all. I knew I wanted to be a singer, even when I was quite young, but I thought I was going to go to univer-sity, study music, maybe be an opera singer, but then it all happened and it was absolutely mental. But no, I didn't have time to "want it" at all.'

Already, some interesting info. Can being famous make you happy if you never *wanted* to be famous? Doesn't that mean you've essentially lost control of your own life? And if you become a major star essentially overnight, there's no putting that genie straight back in the bottle. So, given how she didn't want or expect it, what was being famous actually like for Charlotte?

'It was mad. It was a hell of a ride. The first year or so was amazing, it was *so* exciting. But it wasn't actually the fame that was exciting, it was the opportunities that I had: the travelling, the famous people . . . the *other* famous people that I met. I had my autograph book, and got people in it like Joan Collins. It was all that, just new experiences really.'

From this, it seemed like a life of fame *did* make Charlotte happy, but mostly *indirectly*; she was happier with the consequences of fame, rather than the fame itself. This brought up the subject of reality TV. Viewed through the filter of everything covered in this chapter so far, reality TV suddenly makes a lot more sense; it provides direct and constant satisfaction of those brain bits that seek out gossip and the need to make connections with other people, whomever they may be. I remembered once watching a documentary about the archetypal reality TV show *Big Brother*, where they interviewed unsuccessful applicants who wanted to be on the show. One was a young woman who wanted to be on TV because she 'knew' she'd be famous one day. Only she couldn't give any reason as to *why* she should be famous. She wasn't a performer, she didn't create anything noteworthy; she just existed, and felt that was enough.

In fairness, reality TV means it is indeed possible to be famous for no real reason these days, but this person provides an interesting counterpoint to Charlotte's story. Consider all the things covered in the previous chapter about ambition

and motivation, and our ought and ideal selves. To consciously seek fame is to keep a representation in your brain of your ideal self as famous, which gives us a target, a goal, to achieve and compare our actual selves to. But, fame is very hard to quantify. You can give a precise calculation of someone's height, weight or net worth, but not how famous they are, because it's a far more vague and subjective property.*
Basically, if you set yourself the goal of becoming famous, it's extremely difficult to know for certain how close you're getting to it, especially because you've no idea about what being famous should feel like. Most people have earned money at some point, so there's relevant experience there in getting richer. But fame? Much harder to pin down. It's what philosophers call a sorites paradox, a classic example of which is the question, 'At what point does a pile of sand become a heap?' Likewise, how many times do you have to be recognised on the street, mentioned in a national newspaper or receive fan mail in order to be famous? These things occurring do reveal that our level of fame is increasing, but that's about as much as we can discern. Getting caught in a downpour doesn't mean you know exactly how much rain has landed on you; you just know you're drenched. So, people who want to achieve some unspecified level of fame likely struggle to recognise any progress made towards this goal, introducing failure and uncertainty into their self-assessment. Such things don't make you happy.

By contrast, if you become famous without ever planning to (assuming you've no active objection to it), maybe

---

* Admittedly, with the advent of the internet and social media, where you can keep a precise tally of the number of your viewers/subscribers/likes/downloads, etc., this is possibly changing.

it's essentially tantamount to a lottery win? At least as far as your brain's processing of reward and enjoyment go. You get all the psychological benefits of millions of people liking you, and none of the angst and self-evaluation issues. It's a theory, at least.

However, knowing that people approve of you in the abstract sense is fine, but interacting with them directly seemed to be another matter, as Charlotte pointed out.

'It was fine at first, but when I started entering puberty it got a bit . . . icky. Lots of people were clearly very nervous, even afraid to meet me. Some even thought I was a literal angel.'

This presents another bizarre facet of fame and how it affects our happiness. Maybe it's nice to be liked by hordes of strangers, but as we've seen, the brain craves, and benefits from, social interactions, actual ones occurring between two people. If you're so famous that other people struggle to inter-act with you, that's not ideal. Why would people struggle to talk to a famous person? Well, we've seen how badly the brain deals with even minor social rejection and how much effort goes into avoiding it, so imagine being rejected by someone you really like, and who countless other people approve of and respect/admire. Such a possibility is no doubt deeply unsettling for some people; for the threat-detecting parts of the brain it must be like trying to juggle a live grenade. So of course, they'd be nervous when meeting famous people, with their fearful, risk-averse brains desperately trying to reduce the risk of saying anything upsetting by reducing communi-cation to monosyllabic utterances, grunts or clunky gestures.

Not that it's a painless process for the famous person either. Sitting with us at the table was Charlotte's friend, actor/singer Carys Eleri, and Carys told us about the time she'd seen actor

Rhys Ifans, at the height of his post-*Notting Hill* fame, spend almost thirty minutes trying to cross a room to get to the toilet as people kept stopping him to ask for autographs or pictures. Clearly, many people are confident enough to have no qualms about approaching a famous person, seeing the benefits of meeting them as far outweighing the risk of rejection.

If this happens constantly, it's undoubtedly a big strain on the human brain. In Chapter Two, I described how we need both company and privacy, and that these two seemingly contradictory needs do make sense because social interactions, as enjoyable and necessary as they often are, require effort and energy from the brain, and periods of privacy allow the brain to rest and recuperate. But it's more overarching than that too.

The link between average size of social group and brain size and intelligence was established early on in this chapter. Much of the research into this and the social brain hypothesis in general comes from British anthropologist Robin Dunbar, who coined the term in the first place. Among other things Dunbar has produced, there's Dunbar's number, which is the theoretical maximum number of stable social relationships our brains can sustain at once.[51] We know that social relationships require brain power, and we've only so much to go around. Dunbar argued that the maximum number of relationships we can sustain is 150. While many dispute this, and it's unlikely to be so clear-cut or simple, nobody really argues that there's an upper limit on how much social interaction the brain can sustain, just like there's only so much food you can push into a stomach before it starts becoming harmful.

People who are very famous, whether they want to be or not, will clearly have to meet people and socialise far more often than is the norm, producing much greater strain on the brain (and in Rhys Ifans's case, the bladder). Looked at this way, it's hardly surprising some famous people seem aloof, distant or even curt with fans and well-wishers; it's not necessarily personal or arrogant, it's a desperate attempt to protect their own sanity, wellbeing and happiness. Of course, they could be genuinely unpleasant. That's always possible.

They say 'never meet your heroes', but maybe your heroes aren't thrilled about meeting you either?

I wondered how Charlotte dealt with this side of things.

'I was very good at dealing with people when I was younger, but then between the ages of sixteen and eighteen when I thought I was a proper "bad girl", I was f**ing horrible. I'd be a lot harsher with people. I wouldn't always sign autographs, I wouldn't always pose for photos. But I never really felt comfortable with doing this, it was just an attempt to be cool. I just wanted to impress my peer group.'

I confess this surprised me. I've covered how our brains make us want to conform, to be accepted by those we identify with, but I'd assumed achieving international stardom and being loved by millions would counteract this somewhat. Apparently, I was wrong. Seriously wrong.

'I just learned not to talk about it with my friends; nobody was interested. I'd try and tell them about how I went to the Grammys, where NSYNC sang a song to me, and they'd be like, "Well anyway, guess who was caught necking behind the bike sheds?" They were completely disinterested. Part of that was why I changed the music I was doing,* because I was so isolated from my own age group; my music was not at all what

---

* Charlotte changed from classical to more mainstream pop music around 2005.

most kids my age were into. And that's why I changed it, I suppose, for simple peer approval.'

That Charlotte almost felt a sense of shame about her success, and literally changed the direction of her career, abandoning a style that countless people clearly enjoyed, purely for the potential approval of those she most identified with, shows just how powerful our brain's drive for positive social interactions and approval is. But, as with many other things the brain likes, it must be tangible, something the brain can recognise and appreciate. The smell of your favourite food cooking can be incredibly pleasant, but if you don't ever actually get to eat it, the appeal will soon fade. Similarly, being liked by hordes of unseen strangers is nice, but if the people you actually engage with don't like you, it's not enough to make you happy. The rewards of fame may be many and varied, but the actual direct pleasure and happiness you get from fame itself seems to be rather fleeting.

But it could be that Charlotte's disinterested peer group was a blessing in disguise. Her family and friends are, in her words, 'common as muck', so she was blatantly the only superstar singer among them. To conform, to be part of and accepted by the groups that were clearly most important for her, for her happiness, she had to downplay and ignore the prestige of fame, which very likely proved very useful in the long run. Maybe it's a Welsh thing? We do tend to be very community/family-orientated. And Charlotte's tale of how her father gave her a tremendous telling off when he caught her drinking alcopops underage recalled a very similar tale told by Tom Jones, another massively famous Welsh singer, about waking up from a drunken bender in his LA home in his thirties to be immediately harangued for his obscene behaviour by his visiting mother. Nobody crosses a Welsh mam twice.

Regardless, you can easily see how the sort of downsides Charlotte describes would work against the more 'intoxicating' aspects of fame, and why such things would be harmful in the long run. If media portrayals are anything to go by, when someone becomes properly famous they end up constantly surrounded by agents, minders, assistants, hangers-on, and so on; all people who are dedicated to making them happy. It may sound nice, but it puts them at the direct centre of a social group where everyone agrees that their happiness is the most important thing. They have zero chance of rejection from their social group, so they're missing a major factor in what determines socially acceptable behaviour. It's no wonder mega-famous people seem to occupy a different world to the rest of us; psychologically, they do. Charlotte did divulge a story she was told about an even more famous performer and their ludicrous cocaine-fuelled behaviour. It's utterly hilarious and if I so much as think about repeating it here in print I'll be sued into oblivion before the ink's had a chance to dry.

The point is, maybe fame can make you happy, but it can also mean you end up surrounded by a group in which your approval and acceptance is automatic, not earned. And that can have lasting effects on the brain. Indeed, for child stars it could be even more damaging again. Being famous, as Charlotte keenly observed, makes children different from their peer groups. If they don't make efforts to address this, like she did, they risk being ostracised or excluded. This, and many of the other trappings of fame, can contribute to a sense of isolation, a loss of normal social interactions. And these are children. Remember: what does isolation, according to the available data, do to the developing brain? Disrupts it. *Damages* it. Impairs its ability to be happy! No wonder so many child stars end up with drug, relationship or other serious issues. Exposure to high levels of fame, if not handled correctly, could well be harmful to a child's brain.

Because, as we've seen, our interactions with others are crucial for the wellbeing of our brain, and our happiness. We enjoy positive interactions, we're compelled to seek them out, and forming and maintaining social relationships is a reliable source of happiness. Because we need to conform, we need to belong, in order to feel safe and secure, and for our brains to work like they should. Empathy means we can 'share' emotions, so having other people around when we do things that make us happy just improves the experience for everyone. On the other hand, social rejection is deeply unpleasant, no matter who's doing it. And while it may seem logical that obtaining the approval of millions of others by becoming famous would make us happier, it doesn't quite work like that. It's more the quality of the social approval we get, not the quantity. It's who we get approval from, not how many.

Both pursuing happiness at the exclusion of others and pursuing fame for fame's sake can be described with the same comparison; it's like eating nothing but sugar, spoonful after spoonful after spoonful. It's nice at first, very enjoyable and rewarding, but you eventually end up unfulfilled, behaving strangely and with a lot of people yelling at you.

I admit that in this analogy most of the people yelling at you are probably dentists, but I still stand by it.

# 5

# Love, Lust or Bust

'I recently got to live out a long-term fantasy; a threesome with two guys where I was, shall we say, a "very naughty girl", and they were "teaching me a lesson". But the reality wasn't quite what I'd imagined. At one point I needed to stop to use the toilet, then a delivery man rang the doorbell meaning two of us had to hide behind the sofa, naked.'

Just to be clear, my conversations rarely include tales of multi-partner S&M-infused sexual roleplay. Nonetheless, there I was, sitting in a pub just down the road from London's King's Cross station, hearing about exactly that.

Why? In a way, it's Charlotte Church's fault. Her revealing that mass approval and admiration is no substitute for the acceptance and affection of your loved ones really struck a chord with me. And that term 'loved ones' brought up another major factor in our happiness. It's a common cliché that when you find 'true' love you live happily ever after – indeed, love is all you need, according to the Beatles. Quite a claim, but is it accurate? Is love this omnipresent powerful force that dominates our lives, providing endless happiness? Or a quirk of brain chemistry that's been, ironically, romanticised?

So that's what I decided to investigate next. But one thing was obvious: things were about to get messy. Romantic love, for all that we may think of it as pure and good, has a lot of overlap with lust, that powerful, fundamental drive that compels us to engage in physical intimacy. Sex, basically.

For all that many don't like to talk about it, sex is a major part of most adult human lives, and it affects us in very weird ways. It can make us incredibly happy, euphoric even, or catastrophically unhappy, and the impact it has on our everyday behaviour and thinking cannot be overstated.

However, while I'm not exactly a stranger to either love or sex, I certainly wouldn't claim to be an expert at either, any more that I'd claim to be a film director because I've been to the cinema. So, as usual, I ended up seeking out people more equipped to talk about these matters. One was an expert psychologist and relationship adviser, the other an acclaimed sex blogger and author. As you might expect, it was the latter who regaled me with tales of threesomes gone awry in a public conversation that could easily end up destroying my largely wholesome image.

But still, I'm the neuroscientist here, so before diving into the world of sex and loving relationships, I figured I'd better find out what goes on in the brain when we experience these things. And that's exactly what I did, despite the disturbing effects it was bound to have on my internet search history.

## Too sexy for this book

If we're honest, human sex and sexuality is baffling even before you delve into the scientific literature. We don't need it to survive as individuals, but we spend a ridiculous amount of time and effort trying to get it anyway. Sex features in almost every facet of culture and society, yet it's often considered rude or inappropriate to talk about it. In the UK, the age of sexual consent is sixteen, but you have to be eighteen before you can view explicit material like pornography; so you can *have* sex before you can *look* at it. And despite the eye-wateringly varied range of options humans have when it comes to sexual interaction, those with sexual preferences that don't conform to the 'norm' (i.e. intercourse between a man and a woman) often face stigma and persecution. So, sex can make us happy, but there are many ways in which it makes us unhappy too. Why do we care about it *so much*?

Much of the scientific research into this question focuses on two fundamental components of human sexuality: sexual arousal and sexual desire (aka libido). The former means we are physically and mentally able to have sex, the latter means we want to. And both have considerable effects on our brains.

Arousal is usually the first thing that happens,[1] and is often the result of us perceiving something sexually stimulating. Or, rather, some*one*. Most people are aroused by other humans; specifically their bodies (and, to a certain extent, faces[2]). And while we do appreciate the whole ensemble, certain body parts tend to be more arousing than others. Rippling abs, curvy hips, sensual full lips, large breasts,* firm buttocks, big muscles; these tend to get us more 'fired up' than a glimpse of earlobe or elbow. The reason for this is that they're considered secondary sex characteristics;[3] features that evolved to attract mates, but which aren't part of the reproductive process – much like the large antlers of a moose or tail of a peacock. They're 'sexy', but not 'sex organs', like genitals are. It's believed they imply desirable traits in a potential mate such as fertility, strength and good health. They're basically a

---

* Most female mammals only have enlarged mammary glands when they're lactating, to feed their young. Only human women have to put up with them all their adult lives. Nobody said evolution was considerate.

body's way of putting up a billboard for the instinctive parts of our brains which says, 'Look how fit and healthy I am! My genes must be tip-top. We would make excellent babies!'

Another big factor in arousal is touch. We already know physical contact with another person can be rewarding, but some body parts are particularly responsive to being touched or caressed. The genitals are the obvious ones, as they're densely innervated by nerves that bring about pleasure and reward responses when stimulated, sending neuronal signals to the brain via a number of routes.[4] Genital stimulation is seemingly processed by two parts of the somatosensory cortex; one processes the actual physical sensation, while the other, labelled the secondary somatosensory cortex, adds the 'pleasurable' element.[5]

Interestingly, certain non-genital areas, known as erogenous zones, also provide sexual stimulation when touched.[6] Exactly why ears, nipples, thighs or the neck and so on are erogenous while others aren't remains unclear. Some argue that 'spillover' occurs in the pleasure-processing regions of the cortex, so touching one body part partially activates the (nearby) area of the brain that deals with genital stimulation. In essence, this theory argues that erogenous zones are a bit like hearing your neighbour's music through the thin walls of your home; it's not as loud for you, but you still get the urge to dance. However, studies have found no real evidence for this claim;[7] it could just be that erogenous zones are a quirk of evolution.

But what actually happens in the brain when we experience something arousing? Well, if the cause of arousal is something we see, there's corresponding activity in the extrastriate body area, part of the visual cortex specialising in recognising human body shape and motion, which makes sense. However, there's also activation of the ventromedial prefrontal cortex, which activates, via many important and diverse connections, the other brain regions involved in arousal.[8] If the arousal system is like a fire alarm, the ventromedial prefrontal cortex is the part that sets off the alarm at the first sign of smoke, letting it be known that things are about to get very hot. It also diverts our attention, via the bottom-up system (as mentioned in the previous chapter), to the cause of the arousal.

Once the arousal process is initiated, the amygdala fires up. Being a vital part of emotional processing and learning[9] as well as something of a 'hub' linking numerous important brain regions, the amygdala performs several functions during arousal and sex. One such function is to evaluate the emotional component of the stimulus,[10] to determine whether arousal is 'warranted'. A beautiful man or woman lying naked on your bed? Potentially very arousing. Same person lying naked on an operating table, because you're their surgeon? The amygdala is what (hopefully) would determine that arousal is *not* warranted in this context, despite the similar visual cues.

If the amygdala *does* decide that arousal is appropriate, it sets off various reactions via the numerous pathways and links it has access to. One is the amygdalofugal pathway, connecting the amygdala to the thalamus, hypothalamus, brainstem and nucleus accumbens, and this pathway is supposedly responsible for many of the pleasurable elements of sex and arousal.[11] Another key area triggered by arousal is the hypothalamic-pituitary-gonadal axis,[12] which stimulates and modulates sexual desire via release of sex hormones, namely testosterone from the testes for men and oestrogen from the ovaries for women. And here's where it gets a bit tricky.

The sex hormones are so named because they are released

by the brain during puberty, and are responsible for those substantial and often unsettling changes. Basically, these hormones cause us to develop the secondary sex characteristics, as well as 'activating' our reproductive systems;[13] a crucial part of human development, despite the unseemly hair and bad skin it often results in. But the term 'sex hormones' performs double duty, as they're also involved in sexual activity and arousal. We experience a surge of sex hormones when we're aroused, and there are numerous receptors found throughout the brain that respond to them. Sex hormones make the relevant body parts more sensitive and receptive to contact and sexual activity, which undoubtedly helps increase arousal.[14] But can they be said to *cause* arousal?

Testosterone is the most extensively studied; it's present in both men and women and seems the most clearly linked to arousal.* It's often claimed that raised testosterone levels cause men to become more aroused and focused on sex, but the evidence for this is far from conclusive.[15] Low testosterone can cause erectile dysfunction in men, for example, but subsequently increasing testosterone artificially doesn't seem to fix the problem.[16] Why not?

It's even more confusing for women. Menopausal women undergoing hormone replacement therapy, which involves testosterone, regularly report increased arousal,[17] although sensitivity to testosterone varies quite a lot from woman to woman. Oestrogen, produced by the ovaries, is often considered the female equivalent of testosterone, but its role in sexual arousal is even less clear.[18] Add to this the fact that oestrogen is also found in men, and via various processes

* Despite being present in everyone to varying degrees, it's worth noting that most available studies have focused on heterosexual men, for various reasons.

testosterone can convert to oestrogen (and vice versa), particularly in women, and that there are other precursor substances involved, and the whole thing becomes somewhat baffling. What is beyond doubt, however, is that sex hormones are a key, if confusing, element of the arousal process.

So, once the brain is aroused, it sends our signals to the body via the sex hormones and peripheral nervous system.[19] This causes the tell-tale signs of arousal, namely dilated pupils, flushed cheeks, rapid heart rate, and, of course, the rush of blood to the genitals, causing them to swell and firm up (depending on which genitals you have). Basically, we're ready and able to have sex.

This all describes the underlying, instinctive, physiological aspects, the sort of thing we'd expect to see in most sexual animals (which is nearly all of them). But humans have a bit more 'range' when it comes to arousal. Physical and visual cues may be the bedrock on which our sexual arousal is built, but our bulky, powerful brains can go far beyond those basic stimuli, finding intense stimulation in something as objectively neutral as words on a page, or a spoken discussion.

In addition to this, our brains are so invested in sex and arousal that we can be successfully turned on by things *that have not happened and may never do so*. Sexual fantasies are a big element of human sexuality, and occur with similar frequency in both men and woman. Some studies indicate that our orbitofrontal cortex, part of our frontal lobe that handles many sophisticated functions,[20] is an important region for sexual fantasising. And yet it seems counterproductive to spend valuable brainpower on sexual imaginings that are neither contextually relevant nor remotely likely (in most cases). How can that make us happy? Surely, it'd make us frustrated, distracted and grumpy more than anything?

Apparently not; there's evidence to suggest that regular fantasising of this sort improves focus, attention to detail and memory.[21] There are a lot of different brain regions and processes working together to produce something as detailed and potent as a sexual fantasy. Efficient and reliable communication between disparate brain regions is believed to underpin much of human intelligence, so could it be argued that regular fantasising keeps our brains in peak condition?

Moreover, all this fantasising is also believed to help hone and refine our own sexual behaviour and 'abilities',[22] without having to rely on an actual trial-and-error approach to sex, which would obviously be massively embarrassing. All of us in our daily lives constantly and impulsively think up worst-case scenarios and potential hazards so that we can anticipate them and react accordingly, rather than having to figure things out on the fly when issues do occur. Why wouldn't the same logic apply to sexual scenarios?

As mentioned in the previous chapter, it's argued that our intense sociability is what made humans so smart. But one result of living in immensely large, predominantly peaceful groups, is that we are surrounded by potential sexual partners *all the time*. In this context, it wouldn't be too surprising if the more reactive, instinctive parts of our brains ended up making us think of sex so often. But, whatever the reason, the way the brain deals with it means we can become aroused, ready to have sex, pretty much whenever we want.

That's an important point though; whenever we *want*. Because we don't always want to have sex, even when we are aroused.

## Not tonight, I've got a headache

Many people will at some point experience arousal in places and situations where they have no intention of having sex. I've heard plenty of tales of humiliating physical arousal when experiencing a necessarily intimate medical examination. Many men even report unwanted and confusing erections while sat on a bus, lost in thought.

Much of this happens because several elements of arousal can be triggered purely by reflex, meaning they don't involve the brain at all, but are instead processed by a basic neural connection between genitals and spinal cord.[19] The thrumming vibrations experienced while sitting on a bus may set off these reflexive arousal systems, which process them as a form of intimate touch from an interested partner, rather than the inevitable result of a large vehicle's internal combustion engine. The amygdala may evaluate the context and determine that arousal *isn't* appropriate here, but it's not the only part of us that has a say in the matter. Sometimes we can be 'caught unawares', and the amygdala is left fighting a losing battle against all the underlying physiology behind arousal that's already kicked in, like a lone sailor trying to turn around an oil tanker heading straight for an iceberg of awkward embarrassment.

What this shows us alarmingly clearly is that sexual arousal and sexual desire aren't the same thing; they can, and often do, occur independently.* To understand this distinction, it's

---

* On the flipside, there's also sexual dysfunction, which is often where we want sex but our bodies don't seem to recognise this or respond appropriately, and which can cause a great deal of unhappiness.

important first to consider how desire works at a neurological level.

Sexual desire is mostly processed in the brain's temporal lobe, which makes sense (to a neuroscientist, at least) because much of the limbic system is comprised of temporal lobe areas, especially the amygdala and hippocampus. The limbic system is a complex network of regions that allows emotions and instincts to influence reasoning and thinking, and vice versa. This would clearly be of vital importance when it comes to sexual desire, where a basic animalistic drive determines how we think and act.[23]

The amygdala and hippocampus are highly active during both arousal and desire. The amygdala, as we know, handles the emotional component and determines whether arousal is a valid response. Activation of the hippocampus – the centre of memory processing – maybe explains the flood of arousing memories that can occur when we're in a sexual scenario or why sexual memories are often so vivid and intense. This helps increase and sustain arousal, as well as ensuring we have potentially helpful prior experiences fresh in our mind. Sexual desire also triggers the thalamus, another part of the limbic system, and sort of the Grand Central Station of the brain, which spreads information far and wide.[24] All this means the brain is 'in the mood'.

But emotion and mood aren't enough. The amygdala and associated regions are also linked to the networks vital for motivation, a particularly important one being the anterior cingulate cortex, linked to areas responsible for guiding attention, thinking things through, emotional regulation, and more.[25] The striatum, which compels us to seek out and enjoy interpersonal interactions, also seems to play a key role in emotion and motivation in a specifically sexual scenario.[26] It's hard to imagine anything more 'interpersonal', really.

What this all means is that, while they are separate, arousal and desire are typically intertwined. Luckily, many of the same systems that allow us to experience sexual arousal, desire and the associated motivation at a moment's notice, are also able to 'put the brakes on', so we don't get carried away in fits of uncontrolled lust on an hourly basis.

As mentioned, the amygdala helps evaluate the emotional context. Likewise, the anterior cingulate cortex, while important for sexual motivation, is also crucial for detection of errors or shortfalls in performance, and regulation of appropriate reward. Put simply, it determines whether what we're doing is 'good enough', and motivates us to address this if not. It's no great leap to see how this may be part of the reason why most people don't just want sex, they want to be *good* at it too. Hence 'performance anxiety' is a big factor in sexual dysfunction.[27] This may also explain why we consider some people to be 'out of our league'; given how much energy the brain devotes to self-assessment and image, perhaps the brain registers some people as *too* sexy, so prevents us from pursuing them to avoid failure, criticism and embarrassment, no matter how arousing we might find them?

This seems even more likely when you consider that the most important area for control of our sexual urges is the orbitofrontal cortex,[28] responsible for working out whether an action is likely to result in an overall reward, or punishment. If it's the latter, it suppresses our desire and basically acts as the little voice in your head which says 'you probably shouldn't do that'. And self-control is clearly very important when it comes to sex. Say a sexy person drunkenly and openly flirts with you at a party. But you're married. And so are they. To your best friend, in fact. Your orbitofrontal cortex takes this information, calculates likely long-term consequences, and says, 'It may be an enjoyable experience, but this is a *really* bad idea.'

It doesn't have to be anything so stark; if it's the wrong place, wrong time, wrong person, or you're just too tired, the orbitofrontal cortex recognises this and curtails sexual behaviour. Backing this up are studies showing that men with impairments of this region often demonstrate reckless, risky, hypersexual behaviours,[29] while those with abnormally high activity in the orbitofrontal cortex (for whatever reason) often report sexual dysfunction and reduced libido.[30] Also, complex frontal-lobe regions like the orbitofrontal cortex are among the first parts of the brain to be suppressed or disrupted by alcohol, which explains a lot.

So, while we have all these parts of the brain pushing us towards having sex, some also play a role in holding us back. Intensely pleasurable sex may make us happy in the short term, but our brains are sophisticated enough to realise it isn't always the best idea – showing again how happiness, for us humans, is about much more than instant gratification and pleasures.

## How was it for you?

Nonetheless, once we're actually *having* sex, these parts telling us not to are superfluous. Sex usually requires a sense of abandon, of losing yourself in the moment, so self-analysis and hesitation aren't helpful. Therefore, the orbitofrontal cortex mostly shuts down during sex.[31] In contrast, many other areas are going flat-out, while our aroused body and brain are experiencing the ever-increasing pleasure. Dopamine-based activity in the reward pathway goes into overdrive, sensory signals to the brain from the genitals 'surge', all the other senses are firing with the visceral intensity of it all. When we reach orgasm, our reproductive processes do their thing and we're consumed by a wave of intense pleasure, much like a heroin high. Our cerebellum, important for motor control, is also overstimulated (resulting in all the weird clenching and facial expressions).[32]

This is a broadly accurate overview, but the specifics of what's happening in the brain during orgasm are more uncertain. What data there are suggest that, in terms of the pleasure aspect, men and women experience orgasms in very similar ways,[33] as they share basically the same reward-processing systems. However, some studies suggest that, for women, orgasm leads to 'shutting down' of much of the brain regions involved in sex and emotion, particularly the amygdala, to the extent that they can't feel emotion at all.[34] Arguably this would mean that, during orgasm, women *can't* be happy – or feel any other mood or emotion. But it's not an absence of sensation really, it's like being deafened by a tornado, or blinded by a flash; there's just *too much* sensation, so the brain has to basically hit the circuit breakers to stop things being overloaded. And that may be what's happening with orgasm; for a brief period there's too much activity to handle for the normal apparatus that deals with mood and emotion. One theory as to why such a thing would evolve is that the orgasm phase is most important for reproduction,*

---

* In men at least. There's actually quite an impassioned debate still going on about the exact purpose of the female orgasm, in the evolutionary sense. Some argue it's for enhancing pair-bonding and, indeed, aiding reproduction, while

so the emotion centres shut down to prevent anxiety or apprehension, which would otherwise disrupt the process.

However, later studies showed these same brain areas show *increased* activity during orgasm.[35] How could two very similar studies into the same thing produce wildly different results? One big factor was *how* the orgasm was achieved. In the study where women's brains 'shut down', their partner was stimulating them to orgasm, whereas the increased brain activity was shown in a study where subjects had to do it themselves. One conclusion from this is that our brains process sex with a partner and the DIY approach in starkly different ways.

This makes sense, because despite essentially the same 'end point', they're clearly perceived very differently by our brains; during masturbation, our brains need to do much more 'work', as we're having to *think* a lot more to achieve the level of arousal necessary. The complex processes underlying fantasising are fully engaged; even if using material like pornography or erotica, we still have to imagine the things we're seeing are happening to us, or that we're there, or whatever. We've seen that imagined sexual encounters can arouse us, but some studies even suggest that, like attention, our sexual processes can have a considerable 'top-down' component. This means our conscious, thinking brains aren't just about the anticipation but can control and even induce sexual stimulation. Subjects asked to think about touching their genitals show activity in the somatosensory cortex, as if their genitals *were* being touched, implying an overlap in perception of imagined and real sexual activity.

---

others insist it's just an evolutionary holdover that's present but doesn't have any specific purpose, like male nipples.

Some women even claim to be able to *think* themselves to orgasm,[36] achieving climax without any physical stimulation. Imagine that.

Sounds far-fetched at first, but then eating disorders like anorexia or phenomena like the placebo effect arguably exist because our conscious brain can 'overrule' our underlying biology. Again, why *wouldn't* this be true for sex as well?

Of course, sex with a partner is different. You don't need to imagine having sex with someone if you *are* having sex with someone, therefore those higher brain regions don't have much to do and can be 'powered down'.

So, evidence suggests that, while there are obviously many shared elements, our brains process masturbation and sex differently. This would explain why, even though we humans can trigger our sexual reward systems ourselves, whenever we want (within the bounds of the law), we're seldom content to do *just* that, so still pursue sexual partners. In fact, evidence suggests that, over time, excessive masturbation (for men) plays havoc with sexual wellbeing, significantly reducing libido and ability to become sexually aroused.[37] Luckily, it's not permanent; things go back to normal after a few months of restraint. But that such occurrences haven't been reported in those who engage in particularly frequent sex with partners* suggests that, when it comes to sex, our desires and happiness aren't just based on achieving intense-but-fleeting pleasure.

No, we need somebody. Somebody to love.

---

* This is a tricky area to explore, though, as it's not certain whether sex 'addiction' is a genuine clinical disorder or something subtler, like a combination of several factors. It's not widely recognised by the psychiatric profession, as a result.

## How the brain makes love

At this point I realised I'd inadvertently stumbled on what may be the point where sex crosses over into love, of the romantic sort. It's all well and good describing the pleasure we get from sex in terms of neurochemicals and neurological reward pathways, but it's very telling that the effects of all this on our brains can vary so much depending on whether we're with a partner. It suggests that while we *can* activate the pleasure-inducing sexual reward systems in our brain ourselves, this isn't a great approach when it comes to happiness. It's like skipping to the end of a brilliant novel; yes, you know how it turns out, and it's a lot quicker and easier, but you've missed out on so much.

But love, well, it's complicated, isn't it? Sex is messy and often confusing but at least has recognisable parameters. But love? It's all over the place. It's immensely rewarding, or psychologically devastating. We spend our whole lives searching for it and may never find it, or find it but not realise it until it's too late. We can even be largely indifferent to it, only for it to completely blindside us when we least expect it, like being hit by a juggernaut while making a sandwich. In the kitchen of our eighth-floor flat. And if we do strive for it and find it, despite the assurances of 'happily ever after' or 'till death do us part', and all the time and effort invested, it can still all fall apart.

Basically, love and romantic relationships are confusing. And if I, a full-time nerd with a romantic history about as exciting as a recipe for toast, were to have any hope of understanding them, I was going to need some advice. First, though, I wanted to make sure I knew as much as possible about the fundamentals; in this case about how our brains process love, and the effects it has on them, and therefore us. And what happens in the brain to turn someone into *the* one?

My initial assumption was that there is a deep, fundamental link between love and sex, that while you can have one without the other, they often overlap when it comes to neurological processing. After all, one of the main 'purposes' of having a long-term romantic partner, according to the evolutionary evidence, is to raise children more effectively.[38] Basically, love has its roots in procreation and mating, which obviously influences our sexual behaviours. In the previous chapter we saw that the human ability to form close friendships was likely thanks to evolution detaching and adapting the brain systems that facilitate pair-bonding (monogamy) so they can be applied in the absence of mating, but that doesn't mean the original function has gone away in humans.

By an amazing coincidence, as I sat down to look all this up, I checked Twitter and saw fellow neuroscientist Dr Matthew 'Matt' Wall of Imperial College London sharing his latest study which potentially identifies a whole new sex hormone (called kisspeptin, appropriately enough[39]). I decided to call Dr Wall directly, and get him to explain it. As he puts it: 'Kisspeptin was only discovered about ten years ago, and they initially thought it was important for cancer signalling.* Then they found out it's a sex hormone, and it's really important in puberty.'

Dr Wall explained that none of this hormone-led development of puberty occurs without kisspeptin, which 'sits on top' of the whole process in the brain, kicking it off like a

---

* Cancers are so dangerous because they can influence and alter the activity of other cells and tissues in ways that allow their proliferation, and they often do this via secretion of chemical signals. Kisspeptin was originally named 'metastin', presumably with this function in mind.

minor rock slide that ends up causing an avalanche. Indeed, studies have shown that administering kisspeptin directly into the amygdala (of rats) causes the levels of sex hormones (e.g. testosterone) to rise in the body.

Dr Wall's team was looking into the notion that kisspeptin might be the thing that links emotional and sexual brain responses to the rest of the sexual systems in the body. Such a thing would indeed show a deeply embedded link between love and sex. Dr Wall's group then did their study, the first of its kind in humans, into the activity levels of brain regions strongly implicated in arousal when viewing negative, neutral, sexual or romantic images. He told me that, although there were marked increases in activity in these regions when subjects viewed sexual images rather than neutral or negative ones, 'the best result was for the romantic, couple bonding images'.

While it's still early days in the study of this hormone, the fact that kisspeptin enhances both sexual and romantic processing in the human brain strongly suggests both are fundamentally linked.

There are other neurochemicals implicated in this context, of course. Many studies have shown the crucial roles of hormones/neurotransmitters oxytocin and vasopressin in forming long-term commitments. Chemically very similar, both are synthesised in the hypothalamus and secreted by the pituitary gland. While non-monogamous animals often show levels of these chemicals in their brains equivalent to their more faithful counterparts, monogamous species react to them very differently. In prairie voles,* blocking oxytocin in the brain inhibits

---

* A species widely used for research into this area, as they're monogamous, unlike their very similar (physically and genetically) co-species like montaine voles, so

the usual behaviour of female subjects forming bonds with males they've mated with. Monogamous animals also typically have a much greater density of oxytocin receptors in the nucleus accumbens, and considering that sensory cues associated with partners trigger this release of oxytocin, this would mean a partner's presence causes the experience of pleasure and reward.[40]

If this is also the case for humans, this could explain why we feel so happy and contented when we're in love; we're essentially 'high' (remember, the nucleus accumbens is constantly singled out in drug addiction[41]). Seeing our beloved literally induces pleasure. No wonder we're so fond of them!

Then there's vasopressin, usually considered to be a key factor in long-term mating tendencies, in males specifically. Male prairie voles, and other monogamous species, have more vasopressin receptors in the striato-palladial region, a complex network of areas incorporating things like the amygdala, globus palladus (responsible for movement coordination) and striatum (which includes the nucleus accumbens).[42] There are also many vasopressin neurons that extend from areas like the striatum and amygdala to the forebrain and frontal lobes,[43, 44] usually indicative of a 'direct' role in influencing behaviour.

Somehow, vasopressin action compels males to stick with their partners, something comparatively rare in nature. Supporting this claim is the interesting fact that genes for vasopressin receptors seem 'unstable',[45] meaning the number of vasopressin receptors in the striato-palladial region varies a lot between individual males. And yes, fewer vasopressin receptors in this region does seem to result in

_____

it's possible to compare the brains of the two and spot the subtle differences that would suggest a tendency to commit long term.

reduced tendencies towards forming pair bonds. Basically, if you're less sensitive to vasopressin, you're less able to successfully maintain, or even want to enter into, a long-term relationship. This data comes from vole studies, although there's evidence it applies to humans too,[46] which might suggest some men are biologically averse to long-term relationships. Maybe the tired sitcom cliché of 'commitment phobia' has a genetic basis?

## Love is in the blind eye of the beholder

That lust and love are linked isn't a new observation, of course. Many describe falling in love in terms of a lust–attraction–attachment sequence (or variations on these terms).[47] We start by being aroused by someone we find sexually stimulating. This can then develop into an attraction to one specific individual to the exclusion of others, as opposed to just a generalised arousal response to someone sexually appealing. We think about them constantly, our mind is attracted to them. Finally, assuming we do end up in a stable relationship with them, we become attached to them. The initial dizzying intensity fades, to be replaced by a sense of comfort, satisfaction, security and familiarity with your increasingly long-term partner. The happiness we feel here is more of the contented, relaxed sort.

Exactly what happens when we go from not loving to loving someone is hard to say. You can't put someone in a scanner and say 'OK, fall in love . . . now!' For monogamous animals, and indeed some humans, it may be just a combination of situation, availability and basic physical attraction. An acceptably attractive potential mate is available and open to your advances, you have no obvious alternative options and there's nothing to suggest you'll find any in the near future, so it would be the logical choice to form a long-term bond with this individual.* And most animals form long-term bonds *after* mating, according to the data, so the whole 'no sex before marriage' thing looks to be something of a human creation.

But that itself is both interesting and relevant, because it shows once again how humans and their hefty brains make things like this more sophisticated, and therefore confusing. Just as we can become aroused via our abstract imagined fantasies or simple words on a page, so we can fall in love with someone *we've never met*. How many serious romantic relationships happen online nowadays, between people who aren't even in the same city/country/continent? However many it is, the fact that it can happen at all is an amazing display of the human brain's power. Or, looked at another way, its flaws. It reveals that we don't specifically need a potential partner to meet a checklist of physical features before we're able to fall in love with them.

We've seen how quick and easy it is for our brains to 'connect' with someone else's thanks to our incredibly sociable nature, and it looks like this influences our brain's inclinations to form romantic associations too. Our powerful cortexes and their sensitivity to interpersonal communication means something as simple as an email exchange with someone can reveal all manner of things about them: their sense of humour, their attitudes, their likes and dislikes, their ambitions, and so on. From this, we are usually more than capable of forming a detailed

---

* I first wrote this passage as a message in the first (and, at her request, last) Valentine card I ever gave my wife.

picture of someone in our heads, and if it's one we happen to like a great deal, then why wouldn't we fall in love with them based on something as minor as a text-based conversation?

I did say it was a weakness though, and it can be. Because it means our brains actually 'create' a representation of someone from relatively limited information, meaning it has to do a lot of guesswork and extrapolation. If our brains were 100 per cent logical this might be OK, but they very rarely are. And the human brain is generally optimistic when it comes to things like this; if it's something we want, something we like, our brains are predisposed to make us happy so we interpret or analyse things with a very positive bias.[48] As a result, the image we create of someone from limited data is likely to be more flattering if it's a pleasant, potentially rewarding interaction. Basically, our brains *want* to like them, so we assume they're worth liking, which colours our perception of them a great deal. That's assuming the other person is being completely honest about themselves too, which is rarely the case.

And that's how we get things like 'catfishing',[49] where people create fictional online personas to dupe other people into falling in love with them. Exactly why they do this is a psychological can of worms for another time, but the fact that it's even possible shows how easy it is for the human brain to fall in love with someone. It also shows that, as happy as love may make us, it's often a barrier to thinking logically or rationally, even for our formidable brains. Why would that be the case?

When we fall for someone, and fall hard, studies have revealed there's a substantial increase in central dopamine levels,[50] the neurotransmitter integral for feeling a sense of reward and pleasure, as we know. And what could be more pleasurable than finding the love of our life? But the brain is more sophisticated than that, and dopamine has many different roles. It's necessary for the emotion–motivation processes that guide our actions, and also regulates *anticipation* of reward, meaning we're constantly primed to seek out and achieve the thing that provides a reward. This puts us in a constantly heightened, focused state.[51] A human in love will typically go to great lengths to be around or even just see the object of their affections, and this may explain why.

As well as dopamine, there's a notable increase in noradrenaline in the brain and body when we're in love.[52] This heightens attention, short-term memory and goal-driven behaviour. Noradrenaline, as the name implies,* also influences the release and action of adrenaline, the neurotransmitter/ hormone that triggers the fight-or-flight response, hence people in love can often seem nervous and twitchy. Noradrenaline can also cause sleeplessness, and is particularly involved with heart function,** which would explain why our heart suddenly starts going bananas when we're in love.

As a result of all this, levels of serotonin (the neurotransmitter seemingly vital for feelings of calmness, relaxation and emotional wellbeing) are *reduced* when we're in love, which has potent consequences. Imbalances of serotonin can have substantial effects on our moods,[53] hence modern antidepressants work by increasing neuronal serotonin levels (as covered in Chapter One). Also, we lose sleep, deal with intrusive thoughts,[54] our motivations and drives are altered, meaning

---

* In the US and other places, it's named 'norepinephrine', and adrenaline 'epinephrine'. I refuse to call them that. I will literally fight people over this.
** Perhaps explaining why the heart ended up as the symbol of love, and a central theme of several billion Valentine's cards and other related tat.

things that once gave us pleasure seem inconsequential now, so we end up ignoring our usual friends and pastimes, much to everyone else's annoyance. Such behaviours can also be seen in obsessive compulsive disorder.[55]

If you've ever fallen heavily in love with someone, or been around someone who has, this probably all sounds very familiar. Expressions like 'being crazy' about someone, 'lovesick', or 'head over heels' imply instability, a loss of control and rational behaviour, which does seem to be the case. It's perhaps no wonder that being in love during that scarily intense attraction stage can be so disruptive.

It's not just chemicals though. There does seem to be a specific network of brain regions, featuring familiar areas for emotion and motivation like the putamen, insula and anterior cingulate cortex,[56] that's particularly active during this attraction phase. Interestingly, some studies show activity is *reduced* in areas like the amygdala and posterior cingulate gyrus,[57] key areas for detecting and processing negative stimuli and emotions. These, and other areas responsible for critical thinking and threat detection, are suppressed when we're in love, hence loved-up couples are so damn cheerful all the time and nothing seems to bother them; the parts of their brain responsible for detecting and processing unpleasant things, and the subsequent stresses and concerns they can induce, don't work as well when we're in love. You're simply less capable of worrying about everyday things, so of course being in love makes you happy; your brain is flooded with the chemical responsible for feelings of pleasure and reward, and your ability to experience stress and worry is diminished.

Cynical types need not despair though, as there are downsides to all this. Not least the fact that our ability to think logically about our beloved is substantially reduced. The brain already has optimistic biases for things we like, and if you shut down the fault-finding abilities too, love makes us immune to a person's flaws. Have you ever wondered why people stay with partners who are, to put it mildly, dreadful? It's incredibly infuriating for still-objective friends observing from the sidelines, as it defies all logic and reason, and means they have to watch someone they care about being harmed or exploited. Falling in love with someone is hugely demanding for the brain, and love makes us happy, so, worryingly, our brains go to extreme lengths to keep us loving someone, even if that's logically a very bad idea. Love can indeed be 'blind', after all.

Of course, assuming we do end up with the person we fall in love with, that tumultuous early period passes eventually, and we end up in the 'attachment' stage, hopefully forever. Our brain has adapted to the constant barrage of fluctuating chemicals our infatuation caused and regained some stability; stress chemicals like cortisol recede, and calming serotonin levels go back up.

One of the ways our brains maintain this stability is by forming what is essentially a 'mental model' of how the world works,[58] on which we can base our decisions and expectations in any given scenario. It's formed from all our experiences, memories, attitudes, beliefs, priorities, and so on. And pretty soon, our lover will be a very big part of this; as they become a prominent element in our happy memories and our experiences, our mental model updates to include their constant presence as an underlying factor. The assumption that our partner will always be there is a vital element of our plans, our understanding, our predictions, and so on. Our happiness, therefore, is contingent on their continued presence. Basically, because of how the brain works, if a relationship lasts long enough, our desire to maintain and prolong it becomes somewhat self-fulfilling.

As ever, though, the brain has a few things in place to help this process. Studies show that couples who have been together for decades who state they are still happily in love have activity in the relevant dopamine reward centres of the brain that is basically equivalent to people newly in love,[59] so it seems entirely possible for our brains to keep all the positive, pleasurable associations long-term. Part of this may be due to vasopressin and oxytocin, which are important for maintaining a loving relationship, as well as forming one, as we've seen.

But still, it's easy to see now how sex and love work in the brain, how they're intertwined, and how they make us happy. The brain systems underlying sexual behaviour mean it's easy to become aroused and motivated to seek out sexual partners, because sex is intensely pleasurable and makes us happy. But, if we find a partner we're especially attracted to, and if the connection is strong enough, we end up fixating on them specifically. That's when the brain goes into love mode, and we end up in a prolonged state of happiness; initially very intense although infused with anxiety and irrationality, before levelling off into a calmer, content type of happiness, pretty much for the rest of our lives, because the loving relationship becomes an integral part of our perception of the world. And there we have it, a brain-based explanation of how sex and love make us happy.

Such a shame that it's hilariously wrong.

## Relationship advice

OK, so maybe it's not *wrong* exactly. Everything I'd uncovered about how the human brain processes intimacy and romance was technically correct, insofar as the available evidence suggests. It's just that this neat and tidy explanation of how love and sex work in the brain is clearly woefully inadequate, because it doesn't explain any of the pitfalls and complications of sex or love, like the stigma experienced by people with atypical sexual desires, or the fact that loving relationships can and do break down, causing immense psychological upset. This is when I finally accepted that I was out of my depth, and decided to speak to an agony aunt. Well, what else was I supposed to do?

My go-to agony aunt is Dr Petra Boynton, who provides relationship advice for the readers of numerous publications, including the *Daily Telegraph*. The *Guardian* once described her as 'Britain's first scientific evidence-based agony aunt', as she's also an experienced social psychologist specialising in human sexuality and relationships, as well as being author of *The Research Companion*[60], a practical guide to psychological research. Luckily for me, she agreed to give her incredibly informed perspective about how we humans think about sex and love in the real world.

First, I asked her why people who fall in love don't always live 'happily ever after', like we're so often led to believe they will. Dr Boynton, in the friendly but world-weary tone of one who knows an incredible amount about something but spends a lot of time working with people who stubbornly refuse to admit how little they know, immediately pointed out that the answer to this problem is contained within my question;

the fact that it's what we're *led to believe* should happen. It's not a biological construct so much as a cultural one, underscored by the fact that other cultures don't adhere to this view.

'Some cultures have more formalised, arranged marriages, where the expectation is that you get married, *then* get to know each other. Over time you would hopefully become good friends, and you might find you fall in love or you might not; you might still have great affection, but your priorities may be around having children, and so on. In such cultures the idea of staying together long term is quite different, it's all about maintenance of happiness, and communication, and wellbeing, and input from wider family and so on.'

To those of us in the Western world, raised on a diet of fairy tales, romcoms and will-they-won't-they TV show arcs, the idea that you'd get married before you fall in love, before you even meet – why, that seems ludicrous! And yet, stats suggest that over 50 per cent of all recorded marriages are arranged in some way[61] (largely because they're common in India and, until recently, China, two countries that account for a third of humanity).

So if arranged marriages are a fact of life for a large part of the world's population, it seems obvious that the Western ideal of finding someone by chance, falling in love and then getting married is not necessarily the 'biological default' for humans. Some of us Western types, with our individual rights and freedom of speech and democracy and all that, tend to shudder at the concept of an arranged marriage. We'd never let other people dictate how our relationships will go.

Except, as Dr Boynton pointed out, we do exactly that. All the time. As she puts it, most of us are subject to the idea of a 'relationship escalator'[62] that determines how romantic relationships are supposed to go, with set stages and a vague-but-unavoidable schedule they should conform to. Have you ever asked someone, or been asked, 'Is this relationship going anywhere?' A common enough question, perhaps, but it reveals the subconscious acceptance that a relationship should be heading towards a specific goal, rather than existing for its own sake – which is what all those underlying neurological processes seem to exist for. There's no known 'must move in together within two years' network in the brain. But we've seen how we're capable of adopting long-term ambitions and goals in the working world, and how these can affect our motivation, behaviour and happiness. What's stopping this same process from influencing our romantic relationships too? The answer is, of course, nothing.

There may be a logic to it, and it clearly does work for many, but there are also plenty of downsides, because it means when two people get together they both have a pre-existing notion of where the relationship should end up, what form it's meant to take. But they may not agree on this front. And even if we do fall in love, we still have all the hopes and dreams and plans and ambitions we did before we met our special someone. Unfortunately, though, it's entirely possible to fall in love with someone who, actively or passively, presents an obstacle to these. Our brains are therefore presented with a decision to make; what makes us happier: our relationship, or all our other plans and dreams? When we're still in the 'loved-up' phase, things will likely be heavily in favour of the relationship, but as that passes it's a lot less clear.

You may want to be a successful solicitor or author or whatever, or your dreams may be more of the interpersonal, romantic sort, where you're married with a family and nice house in the country by age thirty-five. Then, you fall in love with someone, but it turns out they will make it much harder for you to achieve your dreams. They have their own career aspirations which aren't compatible with yours: you want to be a master butcher, they're a hardcore vegan who won't eat anything that's even been looked at by a chicken, or they don't want children, or are divorced so can't face being married again, and so on.

This would almost certainly result in some degree of cognitive dissonance: 'I want to get married/be a successful solicitor, etc., but I also want to be with this person I love, who will prevent these things.' In some cases, our brains resolve this by deciding those other things aren't important after all, and it's the person we're with who matters the most. Or, we'll decide it's our goals and dreams that'll make us happier, so we end up thinking 'maybe I don't love this person after all', and we break up the relationship.

The reason, then, that finding love doesn't mean an automatic 'happy ever after' is likely to be the fact that life doesn't *stop* just because we've found someone to spend it with. The brain's mechanism for making us fall in love with someone may be powerful, but it's still not all-consuming, and life is throwing constant changes and upsets at our nice, calm status quo. Some relationships can endure, even be made stronger by such things, but others won't withstand the pressures the world throws at us.

Maybe the brain's method of creating and supporting love made more sense when we were more primitive creatures who spent our much shorter lives in small, limited communities, but that was a long time ago. With our powerful modern cerebrums granting us rich, long, complex inner lives and a similarly complex society to spend them in, a long-term romantic relationship is obviously going to require a lot more effort to sustain, no matter how happy your partner makes you on a one-to-one basis. In the cold light of day, saying finding love will make you 'happy ever after' is like saying the best meal ever will satisfy your hunger for ever; as nice as it is, it *won't*, because that's not how the world works. Neither the brain nor the world is static, fixed in place. What makes you happy today may not make you happy tomorrow, so any relationship, even the most solid, needs time and effort spent on it for it to endure. Luckily, because it's with someone you love, this time and effort can itself be rewarding and make you happy.

And so, to bring it full circle, one way in which a modern relationship is supposed to be sustained is 'in the bedroom'. A healthy and active sex life is seen by many as the corner-stone of a good and lasting relationship in our modern age. But, as Dr Boynton pointed out, this itself could be another cultural creation.

"This idea that you meet someone and have enormous amounts of exciting, erotic and novel experiences until the day you die is a relatively new thing, and what's interesting is it's on the wane. If you look at many millennials, they've got difficult circumstances to deal with due to economic issues; they might have to live with parents or work longer hours so can't go out and socialise, but (maybe as a result) they also seem to recog-nise that sex is only *one* thing that's important. They tend to report having sex much *less* than previous generations.'[63]

Societal attitudes towards sex and its importance in a rela-tionship are more flexible than most realise. The 1960s and 70s saw the 'sexual revolution', with the introduction of the pill, the women's rights movement, recognition of homosexu-ality (officially classed as a mental disorder in the US until the 1970s), and so on. To what extent this was a backlash to the oppressive approach to sexual norms that came before is one for historians and sociologists to discuss, but it certainly paved the way for a society in which sex featured more prominently. But what effect did this have on our happiness?

As well as the relationship escalator, some refer to the sex escalator, which is where our approach to sex is similarly dictated by expectations and social influences. What 'counts' as sex? When people say they're willing to go 'all the way', where are they going? Why do some forms of sexual interaction matter more than others? As well as this, a lot of modern media portrays a full and constantly active sex life as something to aspire to,* as something 'healthy'. Dr Boynton doesn't think this wise.

'Where did this idea that sex is healthy come from? It never used to be,' as she put it. This doesn't mean it's *un*healthy, it just *is*. Maybe it's tantamount to eating; we need to eat, it's essential for good health, but shoving endless fistfuls of cake down your gullet is nice, sure, but isn't 'healthy'. Perhaps it's the same with sex? Everything in moderation, and all that.

The truth is, people vary considerably when it comes to these things, and one of the best things you can do to be happy, as Dr Boynton goes to great lengths to emphasise, is to step back and think about what works for you, what you want and enjoy, not what societal expectations insist you should want or like. Dr Boynton also regularly challenges the notion that sex is a vital part of a relationship. Sure, we can accept that sex is the most intimate and rewarding thing a couple can do together, but it's not the *only* thing. If a couple is having a difficult period, there are many ways to get things back on

---

* Looking at you, *Sex and the City*!

track. You can take up a hobby together or indulge in existing shared interests, go for a nice walk, do some household task you've been putting off, or as Dr Boynton eloquently put it, 'What if you're just kind to each other?'

At the neurochemical level, any positive social interaction causes oxytocin release, and if this enhances existing bonds, then it would strengthen a relationship. It may not be as pleasurable as sex, but it's a lot less effort too. Whenever couples are having problems there seems to be a reflexive assumption that it may be down to issues with their sex lives. It may be, but it doesn't have to be. It's common to think there's 'trouble in the bedroom', but remember there are many other rooms in the house.

Finally, before our conversation ended, Dr Boynton issued a word of caution regarding my own research.

'You know the reason why so many articles and editors in particular are obsessed with hormones and neurology [when it comes to sex]? Because when you're talking about hormones in your brain, you don't have to talk about putting anything inside your vagina.' Or, indeed, any of the other stuff that sex often involves but that we've decreed it impolite to talk about.

Looking back over my work, I realised I too had fallen into this trap. Everything was nice, sterile, family-friendly, and completely devoid of any of the messier aspects of sex. Admittedly, I wanted this book to be available to read for people of all ages, not hidden on the topmost shelf in the dodgy-looking backroom of the bookshop. But, even with all that in mind, could I really say I've learned all I can about sex and love if I maintained a purely objective, academic perspective?

I realised that, no, I couldn't really. So, that's when I decided it was time to talk to Girl on the Net.

## The love-sex balance

Cut to the shadowy corner of a central London bar, where I'm sitting next to an acclaimed sex blogger and author, and admittedly feeling somewhat paranoid about being seen together. Me, the family-friendly married-with-children science writer who does a lot of talks in schools, spotted alongside Girl on the Net (GotN for short), someone whose eye-wateringly extensive and varied sexual exploits are a matter of public record?[64] What would that do for my wholesome image?

Then I realised that it would probably be fine, because GotN maintains her anonymity due to the nature of her work and society's suspicion regarding anyone who is so openly sexual, so nobody would recognise her in any case.

This anonymity means I can't tell you much about GotN, but I will confirm that she's a woman, a surprisingly tall one, with a face and the standard number of limbs. As well as her blogging, she's written an entire book about maintaining a full, active and varied sex life while in an exclusive, long-term relationship.[65] Dr Boynton warned of the dangers of assuming that sex is the most important part of a romantic relationship, particularly if your own sex drive is relatively low. Now, GotN obviously doesn't have that issue, but sex is clearly very important to her, not just for her relationship but pretty much every facet of her life, seeing as it's what she does for a living.* I wondered how she felt about the notion that people are overstating the importance of sex.

---

* To clarify, her income is earned from writing about sex, I don't mean she gets money from sex in the much-older-but-still-illegal-in-this-country manner.

Surprisingly, the prolific sex enthusiast was very much on board with this.

'I never look at my relationship and think "we need to be having more sex" because I've got some idea of how much sex we *should* be having. I look at it and think, "Am I having the amount of sex that makes me happy, that's right for me?" And me *now*, not compared to me in my early twenties or whenever.'

This is an extremely valid point. When we're teenagers, sex is often at the forefront of our minds because we are undergoing sexual maturity, and our body is flooded with sex hormones on a regular basis, causing that confusing-but-potent effect on arousal and desire in the brain. As we age, it tends to cool off a bit, often just because our bodies and brains are getting older, and sex is a very demanding process. For most people it never truly goes away, though. The male sex drive tends to be more consistent on a day-to-day basis, while the female sex drive often dips and increases in tune with the fertility cycle,[66] so you would expect some disparity between male and female partners when it comes to enthusiasm for sex.

But despite her obvious enthusiasm for it, GotN clearly knows the difference between 'as much sex as possible' and 'enough' sex. The former is far more likely to make you frustrated, particularly in a relationship, because unless you're at it all day every day, there's no real upper limit to the amount of sex it's *possible* to have. But as is the case for problematic drunks everywhere, it helps to know when you've had enough.

Further to this, there's also a quality-over-quantity aspect. What arouses people and what they enjoy varies considerably between individuals. And that's when we started talking about GotN's enthusiasm for the more 'aggressive' type of sex, like vigorous spankings and S&M. None of this holds any appeal for me, so I couldn't help but ask how something that's clearly painful can be perceived as pleasurable.

'I think it's the anticipation as much as anything. I've built up this scenario in my head, I'm so worked up, so when it does happen the pain is almost cathartic, the intensity of it feels sort of right.'

While this may well be part of it, a bit of further digging on my part revealed that, especially for women, areas of the brain particularly involved in pain processing, like the peri-aqueductal grey, are very active during sex.[67] There is clearly some logic to this, because sex can easily be painful (indeed, Dr Boynton gets asked about this all the time). It wouldn't be ridiculous for the brain to evolve a system to deal with this, modulating the sensation of pain during sex so it's actually perceived as something more pleasurable. Which, in turn, would certainly explain why so many sexual kinks and behaviours incorporate pain.

If you're like me and such things hold no interest at all, it's hard to relate to what this must be like. But if you've ever eaten and enjoyed spicy food, you're in roughly the same area, given how capsaicin, the chemical that puts the heat into chillies and the like, literally triggers the pain receptors.[68] And yet, some people (i.e. me) still slather every meal with sriracha sauce.

However, while my tolerance for edible spice is reasonably high, I had to admit that this conversation was making me feel all hot and bothered. I wasn't aroused because of the racy content, it just felt awkward and *wrong* to be discussing someone else's sex life so openly and casually in public. I confessed this to GotN, which led to another interesting point, as she thinks that the general expectation for people to be so private and secretive about sex is a big factor in why people can become so unhappy about it.

'If you have an amazing holiday, you can tell everyone about it, and show them your holiday snaps. But if you have an amazing orgy, you . . . can't really do that.'

An amusing comparison perhaps, but something that clearly has a big impact on our happiness, when you consider the workings of the brain and how our happiness depends on being accepted and liked by others. Like it or not, our sex drives and associated turn-ons are a big part of our identity.[69] Hardly surprising when you consider just how much of our brains are involved in them. But, for convoluted and often ancient reasons, sex is rarely something we talk about openly; doing so can genuinely upset people. Just look at the hostility often directed at the very concept of sex education in schools.[70] And when/if it *is* discussed, it tends to be within very narrow parameters, typically focusing on heterosexual intercourse in traditional, monogamous relationships.

What if you aren't specifically interested in that? For one, you may not be heterosexual. Sexual orientation is determined by numerous factors, many of which we aren't even sure about yet, but it's entirely feasible for same-sex (or other iterations) partners to conform to this societal ideal of how coupling and sex 'should' be, consciously or otherwise.

But what if your preferences and tastes regarding sex are focused elsewhere, or encompass a much wider range? The experience of sex has an alarmingly powerful impact on the brain, so we're quick to learn associations related to it, and because people's experiences differ wildly, so our eventual sexual preferences vary substantially. Say you have your first sexual encounter in the back of a car (happens a lot in films, for some reason). The amygdala and hippocampus are firing like mad throughout, so, especially if it's your first time, the event is likely to be effectively burned into your memory. From that point on, you may have a fondness for car-based sex. This might sound far-fetched, but so are most things we suspect about the brain and sex. For instance, there's evidence to suggest that later-life sexual expression is strongly influenced by the style of parenting you received.[71] Or how about one older study into classical conditioning in humans which exposed (straight male) subjects to erotic images while they handled a woman's boot.[72] Over time, the subjects started becoming sexually aroused by boots and other footwear. Basically, psychologists gave a bunch of unsuspecting men a shoe fetish. For science!

The point is, it's very easy to end up with sexual interests that don't conform to the narrow popular definition of 'normal'. So, what do you do then? You're not supposed to talk about it, we're not educated about how sex works in any comprehensive way, and if you own up to having atypical sexual needs you risk social rejection, stigma, even violence. Sometimes, of course, this is understandable; some people have sexual yearnings or predilections that, if acted upon, cause innocent people to come to serious harm. Regardless of how they were obtained, these sexual desires aren't something society can turn a blind eye to.

Having said that, you can still easily end up with sexual preferences that are perfectly harmless among consenting adults, but condemned by wider culture for being too 'unusual'. So the choices are actively suppressing or ignoring your sexual needs (and we've seen how sex and arousal heavily influence motivation), satisfying them but in secretive ways while maintaining an 'acceptable' front (a stressful way to live), or just coming out and admitting everything, risking extreme social rejection and hostility (and, genuine physical dangers aside, we know the brain is extremely sensitive to rejection).

It's perhaps no wonder that those with 'alternative' sexual preferences are far more prone to mental-health problems; because of how society works, it's automatically a far more stressful existence, and this takes its toll.[73]

This would logically suggest that if you were part of a community that *was* totally open about all the different types of sexuality and behaviours, you'd end up being happier overall. GotN has certainly found this to be the case.

'There are so many communities and groups around sex, ones for people with a particular kink or fetish, or for sex writers like me, and so on. It's probably an overgeneralisation, but I've always found everyone to be absolutely lovely. I think it's because everyone's so used to having to explain their kink or whatever that they tend to be a lot more patient and better at communicating.'

This would make sense; without the constant worry of social rejection, if you get to feel normal, then you're likely to be happier and more content, so you probably would be more patient and communicate better. Although there can sometimes be issues when someone who isn't from one of these communities gets together with someone who is.

'A friend of mine is on the kink scene; he hooked up with this girl who was also kinky, but wasn't part of the community, so they're in bed and, without warning, she shoves something up his—'

There followed a report of a sexual interaction so graphically detailed I couldn't even begin to relay it here, but suffice to say that when it was over the varnish on our table was bubbling. But it served to emphasise the importance of communication and openness during sex, because if you just rely on guesswork and wild assumptions about what people like then you're inevitably going to get it wrong some of the time.

A lot of this applies to love and relationships too. GotN, for all her outgoing sexual interests, is in a relationship one would consider 'standard', as in a long-term monogamous one with her boyfriend, whom she lives with. But while sex is obviously a crucial aspect of a relationship for her, and she and her partner have it more often than perhaps most do, this has not resulted in a carefree and simple existence. Reading her book, it's clear that most of the issues that affect her and her boyfriend stem from other aspects of life. Money, starting a family – the usual suspects essentially. Even if everything is fine 'in the bedroom', that doesn't make every other aspect of life just go away.

Viewing this through the lens of her friends who are polyamorous, or in open relationships, or are swingers, and how much suspicion they're regarded with, GotN wishes there were more acceptable relationship models around, so we could see happy, healthy people who don't conform to the standard 'monogamous heterosexual couple who fall in love and get married etc.' types.

Essentially, GotN has figured out the problem of the relationship escalator again, despite never having encountered the term.

## What's love got to do with it?

The literature of neuroscience shows us that sex and love have considerable effects on the brain. We've evolved to seek them out at any available opportunity, and as a result a great deal of our brain is involved in finding them, and experiences immense rewards when we do. Both love and sex even go so far as to alter our cognition and our perception, to maximise the chance of us being happy, at least temporarily. It wouldn't be much of a stretch, therefore, to say that being in a loving relationship that fulfils all your more basic sexual needs is almost certain to make you very happy indeed, because it provides so many rewards.

That's an absolutely ideal scenario, though, and one that works fine in theory, but rarely in practice. Much of this could be said to be the result of the humans and their brains being *too* successful, to the extent that it's backfired on us. Our brains are powerful enough to cause genuine arousal with just our imagination, or fall in love with someone based on what we think they're like from a few pictures and samples of dialogue. But fantasies seldom play out exactly like we imagine (see GotN's threesome story), and the problem with falling in love with someone so easily is that there's no guarantee whatsoever that your affections will be returned. As a million teenage crushes will attest, it's entirely possible to be utterly obsessed with someone who doesn't even know you exist. This can often be a baffling, stressful, even painful experience. Unrequited love is certainly not a source of happiness.

Basically, our powerful intellects mean we have very detailed ideas of what love, sex and relationships should be like, and these shape our behaviours, our motivations, our expectations. A lot of this is backed up and reinforced by (often illogical) societal attitudes and views, and, being the intensely social species we are, we tend to absorb all this and incorporate it into our own ideas and ideals. Sadly, life takes place in the real world, and said real world often doesn't give a damn about your dreams and desires. You can put ample time and effort into pursuing your romantic or sexual interests, only to have it go unrewarded, and we know how much our brains instinctively hate that.

And if you do find love, that doesn't mean life just stops, or the brain mechanisms that got you to this point wither away. The systems underlying attraction and arousal are still there, so it's entirely possible to be 'stimulated' by someone other than your partner, maybe even fall in love with someone else. It's sad, but it happens. A romantic relationship isn't static, because life isn't. Stuff keeps happening, we keep on living, and have to keep on dealing with different things.

This might sound like a bit of an odd comparison, but think of falling in love as like owning a car. You really wanted a car, you've often imagined the model you'd like. Finally, you get one. Maybe it's not exactly what you were expecting, maybe it's better? But regardless, you've got your car now, and you're happy.

Except, just owning a car isn't the point; you don't just put it on your driveway and stare at it, you need to use it, to get about. This thing that makes you happy still has purpose, has function, is active, and so is a loving relationship.

In this context, maybe sex is the fuel for the car? Some need a lot, some need not that much, some need premium, some need the basic stuff, but it's an essential element that keeps things going. Fuel is important, but it isn't the only thing that the car needs; just filling the tank regularly isn't enough to guarantee smooth running. You need to maintain it, patch it up when things go wrong, service it regularly. And so it is with relationships; sex may be an important aspect, but in the long term it's unlikely to be enough to sustain everything by itself. The brain is a vastly complicated and adaptable organ, and it gets used to anything eventually – even sex if it becomes 'predictable' enough.

Overall, it looks like sex and love can, and often do, make us very happy because our brains assign such importance to them, so we find them immensely rewarding. Unfortunately, our neurological and societal sophistication means there are also countless ways and options for it to go badly wrong, making us much less happy overall. It's invariably a constant slog of trial and error before we can be sure what it is we like and be totally comfortable with who we are. But as long as we don't end up constructing a society that has alarmingly restrictive and often confusing rules about sex and relationships, we should all be fine.

Still, you've got to laugh, haven't you?

# 6

# You've Got to Laugh

'You know why comedy is important? Because it's impossible to laugh and be sad at the same time.'

This profound observation was said by Robert Harper. While he may sound like a classical philosopher, he's better known, to British readers at least, as Bobby Ball, one half of veteran comedy-cabaret duo Cannon and Ball, staples of mainstream British TV throughout the seventies and eighties. He made this claim as a guest on comedian Ian Boldsworth's Fubar Radio show. As a huge fan of Ian's, I happened to be listening to this particular episode while nursing a mild hangover on the train home from London, after interviewing Girl on the Net.

It wasn't just the alcohol after-effects that had left me uneasy though; I was also unsettled by what I'd recently discovered. These things that everyone assumes will make us happy, namely sex, love and romance, can make us seriously *un*happy if our brains are too focused on them, at the expense of other things that likely *would* make us happy. Essentially, the pursuit of happiness is often self-defeating. Perhaps this is the root cause of much of the angst and strife that lies at the heart of what it means to be human?

Then it occurred to me; this profound, existential realisation stemmed from a pub conversation with a sex fanatic, centred around a disastrous threesome. I won't lie, the ridiculousness of it all made me laugh, hard. It terrified my fellow train passengers, but I certainly felt better. And then

I heard Bobby Ball in my headphones, making his interesting claim about comedy and happiness, and that got me thinking once again.

Laughter, demonstrably, has a potent effect on our mood; it makes us *happier*, even if briefly. And, perhaps uniquely with regards to things that make us happy, laughter and humour are essentially omnipresent and instantaneous, available in any situation. They even provide the last defence for your happiness when things go horribly wrong. Phrases like, 'Still, you've got to laugh', or 'One day, we'll look back at this and laugh', reveal that, even if everything is crumbling around you, you can still experience happiness if you've got a sense of humour.

Is this right, though? Can such familiar things as laughter and humour be so powerful when it comes to happiness? What is it about humour that affects our brains? And if comedy and laughter can mean instant happiness, why are comedians allegedly so miserable? I decided I needed to find out.

## You've got to laugh . . . no, seriously, you've GOT to laugh!

How does an elephant get down from a tree? It sits on a leaf and waits until autumn.

Not the strongest joke, I grant you, but one that's important for me. It's the first joke I ever learned; one of my earliest memories is of saying it to a room full of relatives, who fell about laughing. I've no idea if they were just humouring a small child, or genuinely thought it was comedy genius (not a lot going on where I grew up). Regardless, I vividly remember being inordinately happy that I'd just made my family laugh like that.

Why do we laugh, though? Because we've heard some brilliant wordplay, or read an amusingly captioned photograph? Because grandma just fell in the pool? Because someone put a pair of pants on the family dog? Because the vicar broke wind during the wedding service? Countless hilarious things in the world, but why do we humans, in response to them, reflexively emit loud, weird noises due to involuntary spasms of the diaphragm with associated contractions of the facial muscles causing smiles? Sure, many emotions cause a corresponding physical response, most typically via facial expressions,[1] or the 'hot flush' of embarrassment. But laughter is loud, prolonged, causes pleasure and related sensations of its own, and can even be debilitating at times. Laughter is not an emotional reaction; it's an emotional *over*reaction. What's that all about? Thankfully, science has some answers.

Firstly, despite what many assume, laughter isn't a uniquely human phenomenon; it's found in other primates like chimpanzees.[2] Their laughter sounds different to ours; less 'ha ha ha', more 'frantically sawing through a wooden plank'. Nonetheless, human and primate laughter have many properties in common, like 'predominantly regular, stable voicing' and 'consistently egressive airflow'. Moreover, via complex acoustic analysis, scientists have determined these different types of laughter diverged from a common type, produced by a common ancestor species, between 10 and 16 million years ago.[3] Far from being uniquely human, laughter is maybe four times older than humanity itself! And it's not just humans and primates; even *rats* laugh. It's incredibly high-pitched and inaudible to humans without special audio equipment, but it's definitely there.[4]

That these nonhuman species exhibit laughter means it's easier to study it. This begs the question though, how do you make a chimp or rat laugh? Wry observations about the ageing alpha male's climbing technique? Do rats find mice, with their stupid oversized ears and teeth, intrinsically funny? Of course not. You want to make an ape or a rat laugh, you just need to tickle them.

That these animals laugh when tickled suggests that the origins of laughter are based on play. Many creatures show playful behaviour, usually in the form of physical rough and tumble. But how do you differentiate playful behaviour of this sort from a genuine physical attack from a rival? Well, laughter, of course. It's argued that laughter evolved to reflexively signal pleasure and acceptance, to say 'this is OK, carry on' when there is clearly no intention to harm. Laughter has been shown to extend the duration of playful interactions,[5] such as tickling. It also explains why we enjoy laughter so much; it means more play, which is beneficial,[6] so we've evolved to experience reward when we laugh.

Laughter, at least the tickling-induced kind, is seemingly handled by a network of deep brain regions, including the amygdala, parts of the thalamus, hypothalamus and lower regions, and key areas of the brainstem. The brainstem, the 'oldest' part of our brain, controls many of our essential but involuntary functions, including muscles that produce facial expressions and breathing patterns. Studies point to the dorsal upper pons, an important brainstem region, as the laughter-coordinating centre,[7] meaning it processes all the neurological activity that leads to the physiological process of laughter.

However, I specified 'tickling-induced' laughter because the cause of the laughter plays a big part in how it's processed by the brain. For instance, while laughter is enjoyable, many people *hate* being tickled, even though it makes them laugh. This is because tickling is weird, scientifically speaking.

Believe it or not, there are two recognised forms of tickling.[8] The first is soft, gentle brushing of the skin, labelled knismesis, which, the theory goes, feels like a (potentially poisonous) insect on our skin. An evolved dislike of such a thing would just be common sense. There's also deliberate, 'forceful' tickling, the type used in laughter research, known as gargalesis (incredibly ticklish people are said to have hyper-gargalesthesia, worth thirty-two points in a game of Scrabble, just so you know). It is a 'friendly' form of touch, so does the usual thing of inducing sensory activity in the somatosensory cortex, whilst also triggering pleasure and reward-related activity, in the anterior cingulate cortex in this case.[9]

So, tickling *can* be pleasurable, and does indeed make us laugh. But, it also induces activity in the hypothalamus and associated areas responsible for the fight-or-flight response.[10] Essentially, gargalesis (in humans) induces a strange mix of amusement and danger. One theory is that it's an evolved reflex to signal submission to someone dominant during playful interactions. The weird laughing-while-recoiling reflex is a way of saying 'You win, I'm fine, but stop!' It's particularly strong when applied to vulnerable, important areas like the soles of our feet, stomach region, armpits or neck. It's easy to imagine our stronger, clumsier ancestors playing boisterously and inadvertently injuring these delicate regions, so a reflex that limits damage without souring relationships would be helpful for a social species.

Of course, not all humans hate tickling; it's one of the very first things human babies laugh at, providing a simple and effective way for parents to bond with little ones. Babies actually start laughing at around three months, before they're able to walk or talk, which again reveals how fundamental and important laughter is. And the things babies laugh at tell us a great deal about how human laughter works. For instance, peek-a-boo, a parent hiding their face only to have it reappear seconds later, reduces babies to hysterics the world over.[11] So, too, do funny expressions, raspberries on the belly, and more. Comedy gold! Some consider these examples of 'proto humour', meaning things which induce laughter by presenting an unexpected change or surprise, but one that is safe, in a familiar social context.[12] Basically, the baby/chimp/rat experiences something unexpected, but quickly realises that it's harmless, or even positive. They've experienced something novel, so learned something new and potentially useful, in the absence of any danger or risk. This is beneficial for a working brain, so pleasurable laughter is experienced as a reward, to encourage this. Does this, then, explain why we laugh?

Not exactly. It's an important aspect, but not the whole story. For instance, most of the things that make us humans laugh, including simpler things just mentioned, don't involve highly physical interactions, or anything that could end up being dangerous, in any way. Also, we can experience rewarding, pleasurable sensations in silence; just ask any teenager with strict parents who's ever brought a 'romantic' partner home late at night. What's the benefit of displaying this pleasure and happiness so 'openly' via laughter?

There's the 'signalling you approve of this interaction' argument, but this is muddied somewhat by the fact that we sometimes laugh when we're *not* amused, so much so that scientists now recognise two distinct types of laughter; Duchenne and Non-Duchenne, named for French neurologist Guillaume Duchenne, and his interesting work into the related process of smiling.[13] Like laughter, we reflexively smile when we're happy. A smile is basically the corners of the mouth being raised, thanks to the zygomaticus major muscle in the face. We have full control over this muscle, so it's easy for us to smile on command. But a real smile, produced by genuine happiness and pleasure, also activates the orbicularis oculi muscle, which raises the cheeks and forms the classic 'crow's feet' around the eyes. This is a genuine Duchenne smile, and conveys real pleasure, because while humans can voluntarily control the mouth muscles easily, the eyes are a bit trickier.

One result of this is that a 'false' smile can be very obvious. If you're at a wedding, and have spent hours being photographed, you're probably tired, aching and fed up; you won't be genuinely happy, meaning your smile looks ever more forced. Saying 'cheese' only activates one set of necessary smile muscles, hence many a false grin is described as 'cheesy'. You're incapable of a Duchenne smile, and so in photos of the happiest day of your life you often look like you're on the verge of snapping and going for someone's throat.

The same principles also apply to laughter. Duchenne laughs are real ones, caused by a genuine emotional experience. Non-Duchenne ones are largely voluntary, or 'false'; laughter we choose to do. That itself is very telling, that we sometimes feel like we *should* laugh even if we're not compelled to. Why is that even a thing?

Much of this comes down to the actual cause, or the source, of the laughter.

[ 188 ]

## Is this a joke to you?

As mentioned, humans are not the only species that laughs. Fair enough. But we do seem to be the only beings capable of *humour*, or 'the quality of being amusing or comic, especially as expressed in literature or speech'. This definition is fine, but seriously undersells how impressive it is. Think about it: you can say/write a sequence of mere words, which induce enjoyable spasms and feelings of happiness in those who see/hear them. That's incredible! Being able to easily reach into someone's brain and alter their mood sounds like something out of science fiction; the sort of thing that would destroy humanity in an episode of *The Outer Limits*. Yet, that's essentially what humour does. How does it have such an effect on our brains?

The most obvious expression of humour is the humble joke. Despite the tired clichés about science and humour being mortal enemies, there have been many experiments and studies that use jokes to investigate the way the brain handles humour, amusement, comedy, etc.[14] And, scientists being scientists, they've produced a detailed and meticulous catalogue of the *types* of jokes that humans recognise.

Firstly, jokes can usually be split into phonetic (language-based jokes that we hear/read), and visual jokes. The most basic example of both is the beloved/notorious pun, the joke type most often utilised in humour studies. Puns can be visual or verbal, and essentially boil down to specific elements conveying different meanings, simultaneously. For instance, 'Why did the golfer wear two pairs of trousers? In case he got a hole in one.' A visual pun, sometimes labelled a 'sight gag', essentially does the same thing, but . . . visually.

For example, you know when an innocuous image has an element that is noticeably phallic, be it a poorly placed table leg, a shadow, someone's elaborate hairstyle, or whatever? It's clearly *not* a penis, but it looks like one, as well as looking like the thing it genuinely is. And that's funny, because sexual things are embarrassing (see Chapter Five) and nobody said we had to be mature all the time.

Then there are semantic jokes, which challenge or break rules of logic and meaning. As with phonetic jokes these can also take visual or verbal form. A verbal example would be: a grouse walks into a bar, and the barman says, 'Hey, we've got a whisky named after you.' The grouse replies, 'What, Kevin?' Ha ha ha, right? But, why? Firstly, there's a brand of whisky called the Famous Grouse; if you don't know that, this joke will make no sense, because that's what the barman is referring to. But the surprisingly communicative grouse misunderstands and replies accordingly, setting up a tension between possible meanings (which is important, as we'll see soon).

For a visual semantic joke, picture a cartoon of a used car lot where the salesman is a literal clown. A surreal image anyway; clowns don't belong in car showrooms. But when you know about the concept of 'clown cars' it has a whole other layer of meaning, and becomes 'funnier'. An appreciation of the meaning, the implications, is what makes the joke 'work', beyond the level of basic visual stimulus.

We also have language-dependent visual humour, which combines words and images. Spend any time on social media lately and you'll be bombarded with memes and comedically labelled photos, where neutral images are captioned with words that impart a whole new hilarious meaning. A grumpy-looking cat presented alongside some appropriately aggressive captions can (and did) spawn millions of memes and several movies (just google 'Grumpy Cat'), thus demonstrating the potency of language-dependent visual humour.

Then there's the fact that visual jokes can be either 'static' (single images, like the printed cartoon of the clown car salesman) or 'dynamic' (video clips or real-life portrayals of comedic scenes and situations). Dynamic visual stimuli often employ scenarios in which unusual behaviour is demonstrated, or events take an unexpected twist. Some experiments show full episodes of comedy shows or performances, incorporating linguistic and verbal elements, which expand the potential range of humour further. You don't specifically *need* language for something visually dynamic to be funny though, hence Charlie Chaplin is a household name.

Here's where it gets tricky; jokes clearly vary considerably in terms of structure and delivery, but do they have anything in common? Is there one key aspect or element that makes something 'humorous', like how a seam of gold can make worthless rock 'valuable'? According to the neurological data, there may well be. It's complicated, though. While laughter can be pinned on the aforementioned brainstem regions, or the supplementary motor areas (which have been shown to induce laughter in studies of epilepsy), humour is more complex. If you take the totality of scanning experiments that have looked at humour processing in the brain, significant levels of activity have been found in (take a deep breath now) the language processing regions in the parietal and frontal lobes, the visual cortex, cortical midline structures including the medial prefrontal cortex, posterior cingulate cortex and precuneus, the superior temporal gyrus (anterior and posterior), superior temporal sulcus, dorsal anterior cingulate cortex, amygdala, hippocampus, and many more besides. An image of the humour-processing areas of the brain looks like a map of the London underground, but even more confusing (if such a thing is possible).

This is pretty much inevitable; jokes or any other expressions of humour contain a lot of information, be it sensory, linguistic or semantic. This must all be 'unpacked' and processed by the brain, via numerous different networks and regions. However, considerable data-crunching points towards a specific process in the brain where everything to do with jokes essentially 'converges', to form a specific system that may well be what 'recognises' humour. This system is composed of regions occupying the junctions between temporal, occipital and parietal lobes, which is the brain equivalent of an airport that connects to three continents, illustrating the wide scope and range of systems that feed into humour.

What this system apparently does, is both detect and resolve incongruity. It is activated when it recognises something as inconsistent with expectations, or the way events or exchanges usually proceed. We know how things *should* work, how they *should* go, but they often don't, and it seems this system is in place in our brains to recognise when this happens. If normality is subverted it means we don't know what's going to happen next, which creates cognitive tension. However, the same system that recognises incongruity (or one very closely linked; it all happens too fast for our best technology to follow at present) then provides a remedy, removing the uncertainty and dispersing the tension. This is a positive thing for the brain, and so we experience a rewarding feeling.

Basically, all this means that thanks to these complex and powerful systems in our brains, humour can be derived from things being surprising, unexpected or 'wrong' in some shape or form, as long as it turns out to be harmless, or even helpful. Consider tickling or other playful behaviour, those 'primitive' sources of laughter. When an animal or baby is tickled, it's an unexpected experience; tickling may be familiar, but it never happens at set times. So, for the briefest of moments, there's uncertainty regarding what it is. It *could* be something dangerous, so that introduces an element of tension, of concern; we know how quick and sensitive the brain is when it comes to anything potentially dangerous. Thankfully, almost immediately we realise it's not something to be concerned about.

The same applies to physical capering, or rough-and-tumble play, or when another member of our group falls spectacularly into some mud; an unusual thing occurs, something incongruous that isn't part of the predictable stream of events in daily life, which produces an immediate sense of tension and uncertainty. But the brain rapidly works out what happened and that there's no immediate danger to this unusual occurrence. Uncertainty is removed, tension is dissipated, and something novel is experienced, all with no inherent risk. All of these things are beneficial for the brain, so an immediate and potent experience of pleasure occurs. It makes us happy.

That's fine for other species, with their straightforward perception of the world. We humans, with our bulbous cerebrums, we have a far more complicated existence involving anticipation and predictions, imagination, complex inferences, beliefs, goals, empathy, sophisticated communication, densely detailed visual perception, and much more. Because our existence involves so much more 'stuff', there's a far greater scope for things to be incongruous, to be 'wrong' in some way. Be it language, imagery, behaviour, or anything else; if our powerful minds impose rules or a predictable structure on it, these can be challenged, or even broken, causing uncertainty. But, if this can be quickly and efficiently resolved, explained away in some form, then the relief and reward is immediate and potent. And that's why we enjoy humour so much, and so often; our brains try to impose sense and order on the world, but there's so much that can thwart this, so we've evolved a system to spot when this happens, and resolve it as soon as we can. And as doing so is beneficial, so we've grown to enjoy it. Again, it makes us happy.

It's a theory anyway, but a reasonably robust one. For instance, scientists investigating jokes often present subjects with similar 'non-jokes', to make sure the brain activity they're looking at isn't just caused by the sensory aspects of the joke. Sometimes the punchline is swapped for a logical statement, like: 'Why did the golfer wear two pairs of trousers? In case he tore one of them.' Others opt for making the punchline even more surreal: 'Why did the golfer wear two pairs of trousers? In case the magic badger that lives in his kneecap ate one.' In the first example, there's no incongruity, nothing out of the ordinary, so no reason to laugh. In the second example, there's certainly incongruity, but no logical resolution, just more confusion. Again no reason to laugh, because the uncertainty remains.[15] Nothing has been achieved, or learned.

Exactly what counts as a 'resolution' for the incongruity is, thankfully, quite flexible. The answer doesn't have to be 100 per cent sensible; as long as our brains can go, 'Ah, I see what's happening here,' that's usually fine, we're happy with a 'pseudo' resolution.[16] The grouse replying 'What, Kevin?' causes incongruity as to the meaning of what the barman said, but the resolution is provided by us realising the grouse understands English and has his own name. Therefore, it works as a joke. It in no way explains how a bird is capable of speech, or why the barman thinks it's normal that an unattended bird has just wandered in, but that's OK. We're aware that this didn't really *happen*, it's a construct for a joke, so any tension introduced by the 'situation' (which would be minimal) dissipates harmlessly, and we're left with the satisfaction of resolving the confusion of the exchange.

This incongruity detecting/resolving system provides the cognitive aspect of humour, but there's also corresponding activity in the mesocorticolimbic dopaminergic areas,[17] those parts of the brain we've covered already that relate to reward, pleasure, and the corresponding positive emotions, specifically happiness. This provides the pleasure and happiness element of humour. What little data there is implies that pleasure and reward experienced via humour is qualitatively different to the sort of pleasure other things give us.[18] One theory for this is that the pleasure we get from humour is coloured by the satisfaction of working out and resolving the incongruity, implying the mental effort put into processing jokes and humour is itself enjoyable, and something we don't get from other sources of pleasure and happiness. We've seen how the brain is 'aware' of how much effort it's put into something, and how that often makes things more rewarding. The act of figuring out a joke or other incongruous occurrence is a key part of what makes humour enjoyable, not just the eventual resolution. The journey is as important as the destination, for all that it's often over and done with in a matter of microseconds.

This system explains many other aspects of humour too. If you have a more intelligent, faster-working brain, you would likely be better at detecting and resolving incongruity, perhaps even anticipating it, so little effort is required for simple examples. This means it takes greater complexity to trigger your humour system, so you're likely to prefer more sophisticated or 'highbrow' jokes, rather than being reduced to hysterics by a man in a dress. It also explains why a joke is never as good when you hear it a second time; the incongruity and resolution have already been detected and resolved, so there's no uncertainty and no effort required to resolve anything, meaning the stimulating elements of the joke are substantially reduced. It also explains why science is often seen as incompatible with humour; science is largely about reducing uncertainty and abnormalities in our understanding of how things work, while these are crucial for humour. You can see why there'd be a culture clash between the two.

So, humour is the result of our brains detecting and resolving incongruity in the various facets of the things we experience. That's good to know. Except there are still unanswered questions. Why are we thirty times more likely to laugh when part of a crowd than when alone?[19] Why does some humour make us feel bad, even outraged? Where does our 'sense of humour' actually come from? Recognising the neurological system at work is far from the full story of the rich human experience of humour. It's like saying a house is made from bricks; technically true, yes, but you can't just get a pile of bricks and call it a home. There are clearly many other elements and influences involved in the finished product, and I needed to understand these before I could truly explain the link between humour and happiness.

It was time to consult an expert.

## Said the actress to the bishop

In the UK, if you want insight about humour and the brain, you speak to Professor Sophie Scott, cognitive neuroscientist at University College London.[20] Responsible for a great deal of research into the workings of humour and laughter in the brain, she's even performed stand-up comedy herself. And, just as I was writing this chapter, she was scheduled to give a talk at Cardiff University Psychology School, so I arranged to meet her in the school refectory beforehand to ask her about her research.

'We don't definitively know what happens in the brain when we laugh, but we know it has a lot of positive effects. There's immediate reduction in adrenaline, and long-term reduction in cortisol, which reduces tension and stress.[21] It also seems to raise pain thresholds.[22] There's even a sort of "exercise high", caused by uptake of endorphins.[23] This isn't as potent as many seem to claim – to say things like "laughing for ten minutes is the equivalent of a five-mile run" is ridiculous. But still, it's there. However, while there are many theories about laughter and amusement, to me the focus on "amusement" is misleading. Laughter has a far more important social role than just expressing pleasure with regards to humour.'

That sounded odd. Wrong, even. How can laughter not be all about humour? But, it seems Professor Scott has copious evidence on her side, and the more I thought about it, the more obvious it became. We originally saw that animals laugh to show recognition and encouragement of playful behaviour. What's that if not a form of communication, of social interaction? We've seen that much of the human brain is dedicated to and influenced by such things, and ample data suggests laughter and humour are further expressions of our brain's innate drive to make nice with others.

For instance, laughter seems to contain a lot of information, which it likely wouldn't do if it were just the meaningless end result of another underlying process. That would be like discovering your farts are in fact high-speed Morse code. Different types of laughter are recognised and processed in different ways by the brain;[24] for instance, laughter induced by tickling seems to engage far more of the conscious attention processes, seeing as how tickling has that bizarre happy/threat combo going on and is induced by physical activity. More 'formal' types of laughter, such as those induced by jokes and humour, trigger other brain regions, many of which are involved in social awareness and processing. In fact, some studies suggest that someone's laugh is particular to them, meaning they can be identified by it. In many cases this isn't difficult (a fellow neuroscience student of mine would genuinely honk like a goose when sufficiently amused), but it's surprisingly common and potent, with at least one study revealing we can identify people by their laugh more easily than we can by their voice.[25]

It's not just laughter that has a strong social element; the same is true of humour too. Among the many areas of the brain active when we process humour, those often utilised are the frontal cortex regions that handle theory of mind.[26] The ability to infer what someone's thinking, to empathise, to 'know' what's going through their head at any given moment, is a vital element of much humour. Someone doing/saying things that clash with their views or thinking, that's a nigh-on bottomless well of incongruity to inspire laughs, one that countless sitcoms have relied upon. Ask yourself; would practical jokes be as funny if we knew the victim was expecting them? The importance of social awareness in humour is underscored by studies revealing that those with a disrupted sense of empathy, due to things like severe social anxiety[27] or autism[28], struggle a lot with humour that requires theory of mind in order to be effective.

Many of the theories about why humour and laughter evolved are based around the social aspect. Some argue laughter is a way of signalling safety and approachability to those in our group, or those we want to interact with. After all, a smiling, laughing person does come across as more approachable than a silent, brooding one. We know that social interactions were very important for early (and modern) humans, but that they can be quite demanding and time-consuming. It's possible laughter and humour evolved to encourage interactions when desired, sort of like traffic lights for interpersonal engagements.

Another theory is that humour and laughter is a way of expressing conflict and aggression, but in safe and socially acceptable ways, so that tensions and animosity are dispelled harmlessly.[29] Say Walter from the office keeps using all your milk from the office fridge. You could challenge him to a fight, but this is risky, for you, Walter, and your odds of remaining employed. Alternatively, you could make a joke about it, maybe suggest Walter is keeping a hungry cat in his desk drawer. Whatever. The point is, you've called him on his behaviour and highlighted the issue, but everyone else (maybe even Walter) gets to laugh and experience enjoyment, so social harmony is maintained. Not a perfect system by any means, but it provides a way of airing conflict that doesn't end in bloodshed. Basically, it keeps everyone happy.

Yet another theory posits that humour is a form of human 'display' behaviour.[30] Maybe the quick resolution of perceived incongruity had important survival value back in the distant past, but now we do it (incredibly often and in many ways) for the same reason that male deer engage in needless battles during mating season; to show others, namely potential partners, that we *can*. Open demonstrations of humour, quick wit and comedic intellect are indicative of a high-functioning brain. 'Look at me,' it says, 'witness the power of my mighty synapses, watch as I create and resolve incongruity on a whim, with no fear or hesitation.' It also induces pleasure and happiness in others. That's obviously going to make someone, hopefully potential sexual partners, more inclined to like you.

It's a two-way process, though. While humour has a huge role in our social interactions, so social factors have considerable impact on our humour and laughter. For instance, countless tired stereotypes suggest that different cultures have specific traits when it comes to humour; Americans don't understand irony, the British are constantly sarcastic, the Japanese have a sadistic streak, Canadians are unfailingly polite, the Germans have no humour whatsoever, and so on. While most of these are utter guff, studies suggest there *is* a cultural influence when it comes to humour;[31] it's just not so blatant and clear-cut as stereotypes would have us believe. This makes sense; if we accept that humour is derived from recognising incongruity, then this depends on our awareness of when something 'isn't right'. And we only know something isn't right thanks to our existing knowledge of how the world works. This is largely dependent on the culture we grow up in. So, if you're from a place where, say, talking about going to the toilet is commonplace, you will have a different reaction to toilet humour than someone from a culture where it's considered impolite or wrong to do so. Neither one is necessarily better nor worse than the other, it just means the reactions will be different.

Another way in which social context affects experience of humour involves our friend the amygdala. It seems to be the part of the brain that decides whether humour and laughter are 'appropriate'. For instance, when someone says 'I could tell you, but then I'd have to kill you', in response to an innocuous request for mundane information like where the photocopier paper is, the polite response is usually to laugh. It's a tired old joke, but it's unlikely any harm was meant by it. However, if the same words are said by a topless, machete-wielding stranger who you just asked why he's in your garage, we probably wouldn't laugh. The same phrase activates the humour system, or doesn't, depending on social context, and it's the amygdala that scans the available information to make this decision.

There's also a lot of *learning* about when laughter is appropriate or warranted that comes via other people. Professor Scott told me that when her son Hector was younger he would always look at her to see if she was laughing before laughing himself. I remembered my own four-year-old son doing exactly this during a best man's speech at a recent wedding. It seems that, while we laugh instinctively, we gradually learn when and where to deploy it by observing others. Indeed, a 2006 study revealed that deaf people laugh at the same points in an interaction as people who can hear.[32] Because laughter is loud and can drown out what's being said – i.e. the very thing that's causing the laughter – people tend to laugh at the end of sentences, or during pauses for breath. Deaf people do the same. This is important, because they communicate, and laugh, by sign language, a *visual* form of communication; there's no issue of obscuring delivery when you laugh, so there's no need to wait for pauses or sentence stops. But, they do anyway, because the rhythm and placement of laughter is learned at a very young age, and goes deep.

Another weird thing we learn is to laugh when we don't feel compelled to. Remember non-Duchenne or 'false' laughter, mentioned earlier? This is laughter that isn't produced in response to a genuine positive emotion, but it's something we all do if we feel laughing is expected, or would improve an uncomfortable situation. Say your boss tells a particularly groanworthy joke in a meeting, or an acquaintance regales you with an anecdote at a party that's a lot less funny than they think it is. In these situations, you don't want to laugh, but it is expected, and not doing so could make the situation awkward or tense, which you want to avoid. So, you do laugh, but it's a non-Duchenne style laugh. Nonetheless it maintains harmony, acknowledges attempts at humour, and can ensure that you remain an acceptable member of the group. Far from being an annoying habit of sycophants and the sarcastic (although it can indeed be these things), false laughter is an essential behaviour that ensures social harmony and acceptance, keeping ourselves and everyone else happy. Studies show that even chimps do it, for the same reasons.[33]

And there we have an explanation – or, more accurately, several explanations – for the ways in which humour, laughter and happiness are linked. Humour helps us resolve potential anomalies in the things we experience, and this is helpful for the brain, so it's evolved to reward us when it happens. As a result, because they're essentially *dependent* on noticing irregularities or 'wrongness' in our world, humour and laughter are still available to us at the worst of times, when everything's 'gone wrong'. As long as our brains are intact, we can still be made happy by humour, even if only briefly.

But so potent are humour and laughter, and so social a species are we humans, that they've grown and evolved to fulfil more social functions. We can now *create* incongruities, aka jokes, and resolve them at will, displaying our humorous prowess like a peacock displaying its tail. Humour provides constant positive reinforcement in interpersonal exchanges, makes us more attractive to others (within reason[34]), offers a safe way to defuse tension and conflict, and encourages and rewards group harmony. No wonder laughter is contagious and we're far more likely to laugh as part of a group than alone; that's largely *what it's for*! To spread positivity and harmony between people. And all these things contribute to us being happy.

So, logically, to be happy, you should spend as much time as possible using humour and inspiring laughter in those around you. That's a sure-fire way to stay as happy as possible, right?

Right?

## Send in the clowns

If we accept the above argument, then people who work in comedy for a living should logically be far happier than the average person. Stand-up comedians in particular; comedy writing and other behind-the-scenes work is no doubt rewarding, but performing comedy live for an audience means there's no separation between your humour and the laughter it generates, so stand-ups get the whole package as far as the brain is concerned. They should be as happy as can be. Yet, conventional wisdom suggests the opposite is the case, hence the 'tears of a clown' cliché, which implies that most comedians and comedy performers are hiding, behind the laughter, deep-seated misery and pain.

Is this true? If so, why? Are comedians inherently sad? Or does prolonged use of laughter and humour *make you unhappy*? Salt is nice in small doses, but consuming large quantities plays havoc with your health; is this also true for humour and our brains? This was an important issue to address. So, given the relative dearth of studies focusing on comedians and their work, I opted to go direct to the source, and basically ask some clowns if, or why, they cry. Metaphorically, of course. And that's where my friend Wes Packer first came into the frame.

Wes is a stand-up comedian. In 2006 he won the prestigious So You Think You're Funny contest at the Edinburgh Fringe, meaning he got to perform at 'Just for Laughs' in Montreal, the world's biggest comedy festival, after barely a year as a stand-up. Wes seemed destined for stardom. However, this is probably the first you've heard of him, so he's clearly not a household name. What happened? We're good friends, so I figured I'd just ask him direct.

Wes and I both started comedy on the Cardiff scene; I was actually the act on after his barnstorming debut set, and did horrifically badly. Like myself, Wes was born and raised in a South Wales mining valley. Wes could also be described as an 'angry' comic; he specialises in furious but meticulous rants about his own failings and those of the wider world. Does that on-stage rage reflect a genuine dissatisfaction in his life overall? Did this lead to him doing comedy?

'I think I saw comedy as an escape. Us downtrodden work-ing-class valley boys, our prospects aren't brilliant, and I could see my life panning out, full of tedious websites in some horrific office.* I didn't want that. Comedy looked like a fun way out.'

To this end, Wes went all out to make a name for himself, doing every possible gig. But this turned out to be a bad strat-egy. During his 'week of hell', he drove to four gigs on four consecutive evenings, driving over 1,500 miles in total, aver-aging three hours sleep a night, because *he still had to turn up for work each morning.*

'I got in at 5 a.m. on Saturday morning, and at 9 a.m. I was sat in the car outside the mechanics while my wife was inside, and I . . . wasn't right. The sunshine was giving me headaches, I was photosensitive, and every car that drove past made me physically flinch. As soon as we got back home I went back to sleep. Set my alarm for 1 p.m., woke up at 5 p.m. I was meant to be at a gig in London in an hour's time. Didn't make it, obviously. Ended up calling them and saying my car had had a breakdown.'

Swap 'car' for 'mind' and that's a reasonable excuse. But it can't all be pinned on comedy. Wes was diagnosed with depression and anxiety in 2012, but is confident he's been dealing with these issues from a much younger age, which he feels explains his crippling fear of 'losing control' of his anger (a recognised manifestation of depression[35]) to his reflexive tendency to tell jokes in any social interaction, to avoid being more honest and open, for fear of where that may end up. Wes says comedy became almost therapeutic, with the brief intervals on stage 'keeping him going' during the

---

* Wes was a software engineer.

bleaker times. Who's to say other comics don't get into it via this 'self-medicating' route?

Unfortunately, even the success of getting to perform at huge, celebrated venues to large crowds wasn't all positive.

'I don't think I've ever been happier than when I was on stage at Montreal, being played on by a house band blasting out Tom Jones. But afterwards, you go home, go to work Monday, and get called in by your boss for a stern word about a minor spreadsheet error you made last month, and he's telling you everything you did wrong and I'm like . . . I was in Canada last week, being cheered by hundreds of people; I'm sure your Excel files are important to you, but I'm struggling to give a f**k.'

Sadly, Wes gave up stand-up in 2008, reluctantly choosing to focus on his day job. He made a successful comeback in 2011, but withdrew again eighteen months later when the anxiety and depression returned. Now, in 2017, divorced and with nothing to lose (as he puts it), he's trying again. Hopefully third time's a charm.

Clearly, Wes endured a lot of stress from stand-up comedy, but much of that can be pinned on the stressful nature of the job when you're trying to succeed. But what if you *do* succeed? What happens then?

To see if 'hitting the comedy big time' made a difference to your happiness, I spoke to Rhod Gilbert, internationally renowned comedian, star of TV and radio, award winner (including 'Wales's sexiest man', 2010[36]), and, most importantly, the best-known comedian in my phone's contact list. I ended up meeting Rhod in London, at a pub near his home. Like Wes and myself, Rhod is also from Wales (Camarthenshire), and he first attempted comedy after being nagged to for years by his then-girlfriend. At the time he was thirty-three, and a dedicated market research director on the verge of buying the company he worked for. But at the last minute he pulled out of this arrangement, gave up his job, and opted to become a full-time comedian instead, despite the massive reduction in pay.

Although it's tempting to assume an 'escape' motivation again, Rhod denies he was unhappy in his previous job. The reason for his sudden and alarming career change was simple: he got bored. He'd grown tired of market research after ten years, so turned to comedy, something he enjoyed and was clearly good at. Despite never having any goals beyond being able to feed himself, Rhod certainly succeeded, and is on TV and radio quite often. At the time of writing, though, he's not done stand-up for five years. Because, again, he got tired of it.

'On my last tour I did 127 shows over eight months, two and a half hours of comedy every night. I used up all my new material on this show, so at the end of it I had to start preparing for the next one, and I was staring a blank page. I haven't had a blank page in ten years, and that's the hardest place to be. I just couldn't bring myself to do it all again.'

It seems, if Rhod is anything to go by, that even the innate pleasure of making large groups of people laugh, the result of deep, fundamental neurological processes, can wear off with prolonged exposure. Habituation strikes again! Still, is there anything besides that about being a successful comic that could make you actively *un*happy? Apparently, yes.

'When you're starting out in comedy, largely the world is supportive. People encourage you, they find things to praise about your set. But once you get past a certain point of success, and other people like you, that's when other people come out and say they don't. Actually, it's more that up to a certain level they don't like what you're doing, but after a certain level they don't like *you*.'

So, being a famous comedian means you're more likely to be constantly facing criticism, not less. And that's not nice. Laughter is, as we've seen, an inherently social act, one 'intended' to obtain approval and acceptance from others. But if you get to the point where your attempts at eliciting laughter end up with you being condemned and abused by strangers, surely that's not pleasant? We know the human brain is extremely sensitive to even minor rejection, but to get it from countless strangers when you're just trying to make them chuckle? Rhod freely admits that it genuinely upsets him, and he's not bothered about 'toughening up' or anything like that; he just does what he can to avoid it, like not getting involved with social media.

But despite this and his curmudgeonly on-stage persona, is Rhod actually happy with his success? He says yes, he's a happy person by default, but his happiness is like plate spinning, trying to keep several in the air at once.

'This plate represents my career, this one my family, that one's my finances, and so on. I keep them all spinning. If one wobbles, I focus on that; if another wobbles, I move on. I keep them all going, then I'm happy. And I don't take on any more than I can handle, hence not doing all that social media stuff. I've not got the time or patience to spend twenty-four hours a day interacting with strangers who want to tell me why they hate me.'

What does Rhod's decision to step away from stand-up tell us about humour and laughter, with regard to happiness? It suggests that, as omnipresent and powerful as it may be, humour still has limits. Evidence suggests that incongruity, violation of norms, is essential for humour to be effective, and maybe this is true for those providing it as well as those experiencing it? Monotony, familiarity, these all bring about predictability and reduce novelty, which we know reduces the pleasure it's possible to take from the things in question.[37] It's even possible that the fundamental process of habituation, over time, wears away at the positive effects of doing comedy. In that situation, the downsides of a life of comedy could start to gain the upper hand, and make it more a chore than a pleasure, at the very least.

Rhod hasn't *quit* though, he's just taken a rather long hiatus (five years) from performing stand-up, and since meeting him he's made some tentative returns. He says it got to the point where performing became 'a relief, more than a joy', and doing the gigs was a necessity rather than a pleasure; he effectively 'lost the buzz', so stepped back. But his underlying enthusiasm for it remained, and has returned to the fore.

So, what does it take for a comedian to say 'no, that's enough now'? To answer this, I ended up driving to a remote farmhouse somewhere in the middle of rural England. It may sound like the set-up to a slasher film, but rest assured, this doesn't end with my mangled remains being buried beneath a disused barn. The farmhouse was the home of Ian Boldsworth, the comedian and radio presenter whose interview with Bobby Ball inspired me to look into humour and happiness in the first place. It's also Ian who got me into blogging after I read his captivating blog about his 2006 Edinburgh run. He then got started in podcasting, which I listened to during the long boring hours in the lab while I was completing my PhD. Without Ian, you certainly wouldn't be reading this now. Meeting with him was a bit like Luke Skywalker meeting Obi Wan Kenobi, if Obi Wan only got Luke into Jedi training by accident.

Ian, a barrel-chested, long-haired, bearded Northerner (imagine a Viking that's taken up skateboarding), is typically and unflinchingly honest and open, even when it comes to his own issues with mental health; he deals with regular bouts of severe depression, and has even done an Edinburgh show where he tells the story of a suicide attempt (2014's 'Here Comes Trouble'). And, most relevantly, he's recently 'stepped back' from performing stand-up comedy, limiting himself to one gig a month. I wanted to know why.

'I just stopped enjoying it,' Ian explained 'I don't think I've ever really thought of stand-up as "fun". There have been moments in it, like when you say something you think is funny, and the audience really likes it and you're all having a good time together. And when I was in a double act, or touring with a fellow comic, those times felt like a laugh, like dicking about on stage. I enjoyed those. But overall, I've always felt a bit ambivalent about it as a thing.'

This general ambivalence obviously contributed to Ian's decision to step away from stand-up. There were many other reasons, but they all essentially boiled down to the fact that comedy was becoming too restrictive, incorporating too much of the workplace culture, like rules and career goals, that Ian and others like Wes were so keen to get away from. I've heard similar tales from other comics.

Disillusioned with the live scene, Ian now focuses more on his broadcasting and podcast output. His own dealings with mental health led him to make the award-winning 'Mental Podcast',[38] full of candid and illuminating discussions with those dealing with their own issues and disorders. He's also just finished 'The Parapod', where he argues about ghosts, mysteries and conspiracies with fellow comedian and enthusiastic believer in all things supernatural, Barry Dodds.[39] Ian is undeniably very happy with his lot at present, doing his own thing on his terms, as and how he wants to. This would address the issues that Rhod observed, namely the lack of novelty and too many people knowing about you and bombarding you with criticism as a result. If you keep doing new things, and shift your focus before anything becomes too 'big' or predictable, maybe that can keep you happy? But, perhaps most importantly for my investigation, Ian's work doesn't now hinge on making people laugh. It may seem odd, but perhaps when you spend so long experiencing humour and laughter, it becomes easier to be happy without it? People often seem surprised by how serious or 'normal' comedians can seem in conversation, that they aren't their thirty-joke-a-minute personas they present on the stage.

Maybe laughter and humour make you happy like money does; very potent up to a point, but once you experience 'enough' it starts to seem less important? Ian even semi-guiltily admits to nowadays experiencing a similar buzz of satisfaction when he gets angry, irate responses to something he's created, as when he gets praise.

'It's weird, but when people get angry enough to object to something I've made, I'm like "good", because they're clearly not my target audience, so I'm obviously doing something right.' That comedy can eventually turn rejection into a *positive*? That's very incongruous. Which is appropriate, given how we know it works in the brain.

## When the laughter stops

As fun as it was talking to my comedy friends and idols, I still had to sit down and work out what, if anything, I'd learned about how constant use of and exposure to humour and laughter affects us. Does it make you unhappy, or not? And in either case, why? Well, there are several points worth considering.

Firstly, performing comedy (successfully) is immensely stimulating and pleasurable. Remember, our brains experience reward, a little burst of happiness, in response to any positive social interaction, and this is even more potent if we make someone laugh. Laughter signals social acceptance, approval, group harmony – all things our brains like. So, to make a room (or an arena) full of people laugh, that'll trigger the brain's reward circuits like nobody's business. Professor Scott and many others report extreme giddiness and trembling after successful comedy sets. You can see how someone could become 'hooked' on the feeling.

During our chat, Wes astutely compared it to drug addiction; the satisfaction and pleasure obtained while on stage was often considered worth the constant effort and hassle surrounding it (driving long distances, sleep loss, the self-control needed to interact with arrogant hecklers and clueless promoters without stabbing them directly in the eyeballs, etc.) just like how addicts are willing to endure the associated dangers of drug use to sustain their habit. Is this a fair comparison? Are comedians essentially junkies for concentrated social approval? That's probably a bit extreme, but that doesn't mean it's an idea entirely without merit.

Remember, drug addiction causes shifts in your actual

thinking and motivation[40] due to the constant and intense stimulation of the reward pathway literally altering the connections between it and the frontal lobe regions responsible for cognition, restraint, and so forth. As a result, addicts become focused on their vice of choice, to the detriment of all else, including interpersonal relationships, hygiene and obedience of the law. I'm not saying comedians are humour junkies, who all end up slouched on filthy mattresses, telling jokes in crack dens (or 'craic' dens, if you're Irish?), but presumably the same underlying neurological systems would be present in their brains as anyone else's. So, if you get copious social approval and validation on stage several times a week, presumably that would satisfy any instinctive need for longing and acceptance, so you wouldn't need or want to get it elsewhere. Compared to crowds of people laughing at your words and applauding your name, a 'meets expectations' rating in your bi-annual performance review is going to seem pretty feeble. Remember, that's exactly what happened to Wes.

Why would performing comedy be so stimulating though? We all laugh and joke with people all the time, but it's rare to experience such an intense 'high' after a meaningful conversation, or even a laughter-filled evening with friends. What's different about doing comedy? The answer is a sense of *risk*. An audience of people approving of you is seriously enjoyable, but they can also *not* laugh; they can reject you, which is a hugely unpleasant experience (trust me on this). We've seen how badly the human brain handles social rejection. In fact, the most common type of phobias are social phobias,[41] meaning people are intrinsically terrified of scenarios where others may end up rejecting them. This logically implies that people fear social rejection more than things like snakes and spiders.

That's how powerful it is.* An audience being unresponsive to, or even actively booing, your attempts to make them laugh is as potent a form of rejection as you can hope for, outside of a romantic partner breaking up with you out of the blue (trust me on this as well). This probably explains why people typically react with more shock, awe and horror when I say I do stand-up comedy than to anything else I tell them about myself. Remember; I'm a doctor of neuroscience who *used to cut up corpses for a living*!

So, if you step on stage to perform stand-up, as far as our brain is concerned, it's basically the social-interaction equivalent of a bungee jump; you know it'll be fine on a conscious level, that no physical harm will come to you, but every evolved survival instinct is screaming at you not to do it, meaning your fight-or-flight system is on high alert. So, if you do well, not only do you have the rewarding feeling of approval and the associated pleasure of that, you also have the titanic relief of having avoided risk.[42] No wonder a good comedy gig can (allegedly) induce a sense of euphoria, and why a bad gig is invariably referred to as a 'death', as in 'Man, you died up there'; it's not literally as bad as death, but it often *feels* like it.

If social rejection is such a big risk with stand-up comedy, then it's more likely to attract those who aren't as bothered by the possibility. So either people with an unshakeable sense of self-confidence (very well represented in comedy, believe me), or people who are desensitised to social rejection, because *they're used to it*. The misfits, the oddballs,

* Actually, this could just mean people appreciate that social rejection is a more likely occurrence, which is fair. In modern society, you are indeed more likely to make a fool of yourself in a presentation than be set upon by an enraged tarantula.

the outsiders, those who don't fit in with 'normal' society for reasons of upbringing, personality or, indeed, mental-health issues. The nature of live comedy is almost self-selecting for those with emotional or related problems that would lead them to often be shunned by the populace, so you'd expect to see them more often in the stand-up world. Going by anecdotal evidence, it seems you certainly do.

The act of performing comedy, for all that it's about humour and jokes and fun, could conceivably exacerbate existing emotional instability. Both Ian and Wes, who have their own mental-health concerns to deal with, described doing routines that meant putting themselves into very negative emotional states; Ian with the retelling of his suicidal episode, Wes with his constant angry outbursts. The human brain is very sensitive to all things humour- and laughter-related,[43] and very adept at reading and inferring other people's emotions. Thanks to theory of mind and empathy and all that, to successfully perform comedy usually requires *authenticity*. An audience needs to believe a performer is being genuine, at least to a degree. So, if your act contains elements of anger or sadness or other negative emotions, unless you're a fantastic actor the only way to convey these effectively is to genuinely experience them, by dredging up relevant memories or putting yourself in that mindset. Essentially, comedians can end up in a situation where they're being rewarded, both financially and with audience laughter and approval, for being unhappy. Wes said it got to the point where he was looking for things to be angry about in his daily life, so he could talk about them onstage. If this is true for many other comedians, it means, thanks to humour and laughter, they're *being conditioned to be unhappy*.[44] That's probably not great, overall.

It seems there are indeed brain-based factors that could explain why comedians, people who specialise in using humour and spreading laughter, would be more likely to end up unhappy than happy, despite how counterintuitive that may seem. Many aren't, of course; a lot of comedians are perfectly happy, and revel in what they do. But, assuming the 'tears of a clown' cliché has to come from somewhere, we can now put together a plausible neurological mechanism to explain it. Performing comedy means considerable approval and validation from others, but risk of considerable rejection too. As a result, those who are desensitised to social rejection would be more likely to attempt it, and humour relying on a sense of incongruity means those with an 'alternative' worldview would perhaps have greater chance of success at performing it. But it also means regularly 'baring your soul' for the approval of others, and if they reward you displaying negative emotions in the name of comedy, it could risk encouraging and perpetuating unhappiness. That, coupled with the nature of the work, could provide a considerable strain on the happiness and wellbeing of those who engage in it, particularly for those with brains and minds that are already inherently vulnerable.[45]

But what does this reveal about laughter, humour and happiness? Well, laughter and humour certainly seem to be powerful components when it comes to making us happy. They are ever-present, versatile, easily deployed and instantly effective, and with many tangible benefits, like enhancing social cohesion, safely releasing tensions and aggression, and even having a protective effect on our ability to withstand and get over stress and trauma. However, judging by what many comedians have told me, it seems laughter and humour are powerful enough to reward, and therefore encourage, unpleasant and detrimental behaviours, which can and do prove harmful in the long term. And, if Rhod and Ian's examples are anything to go by, our brains can become desensitised to laughter and humour if we're exposed to them for long enough.

Overall, it seems that Bobby Ball was essentially right when he said it's impossible to laugh and be sad at the same time (presumably this doesn't include non-Duchenne laughter). But only in terms of that very moment. It looks like laughter and the processes underpinning it do indeed suppress or block out other, more negative emotions as it's happening.[46] And yet it's still rather transient. As well as this, humour is derived from recognising something is incongruous, or 'wrong', but for this to work there needs to be something, be it a rule, norm or expectation, to be 'violated' in the first place. Similarly, laughter may have more of a social role, but it's an enhancing, facilitative one; laughter is more about strengthening social bonds, than creating them outright (although this isn't an iron-clad rule).

Basically, it's difficult for laughter and humour to exist in isolation; they need to have something to respond to, something to be based on. In a way, they are to happiness what spices or condiments are to a meal. The right amount of spice can greatly enhance a meal, or even save a poor one. And even the worst culinary slop can often be salvaged by slathering it with ketchup or salt. And so it is with laughter and humour; they can make an enjoyable situation better, make a bad one OK, and even provide glimmers of happiness when it all goes horribly wrong.

Maybe a different culinary metaphor is needed? Maybe humour and laughter are like the icing on the cake of happiness, so happiness based solely on laughter and humour is like the icing without the cake. It may look pleasant, it might look like an actual cake, it may even taste nice, but it's fragile and unsatisfying – too much of it and it may turn unpleasant, and it doesn't take much for the whole thing to collapse. And because humour seems dependent on a sense of incongruousness, of subjectivity, of unpredictability and surprise, then it seems any attempt to formalise it, to establish rules and structure, to make it reliable and manageable, risks eroding the very properties that make it worthwhile. And how's that meant to make you happy?

Ian summed it up best, with a tale of when he and his then writing partner were brought to the comedy department of the BBC to help write jokes for an upcoming TV show.

'We were just happy to be there at first. We were put in this office that used to belong to [British comedy duo] French and Saunders, and we couldn't help mucking about, taking our photos with all the awards. Eventually, we realised we should get on with some work, and sat down to write. We were coming out with stuff, and it was making us really laugh. I don't think any of it was ever used, but we were enjoying it. So we were laughing away, and then the head of the programme, a well-known comedian in his own right, sticks his head through the door and says, "Lads, can you keep the noise down, we're trying to do some work here." And that should have been a warning sign to me; we were in the heart of BBC comedy, people should have been able to hear the laughter on the outskirts of London!'

That sums it up nicely, I feel. If you dedicate too much time and attention to humour and comedy, to the extent that it becomes your sole focus, it can get to the point where laughter is frowned upon.

Isn't that funny?

# The Dark Side of Happiness

I was a cheerleader once. Not something most other balding thirtysomething male neuroscientists can claim, I'd wager. It was when I was a teenager. A fundraising event my parents organised featured a spoof of WWF-style wrestling,* and I was one of the 'bad guy' cheerleaders. It sounds a ridiculous sight, a chubby teenage boy in a gold wig and black skirt, waving pom-poms around, and it was. That was the point.

Aside from occasionally waking up in a cold sweat and screaming, as an adult I rarely even think about my stint as a cheerleader. It just doesn't fit my present-day image, so it rarely comes up. I mention it now though, because we all have weird, perhaps regrettable things in our past that we'd rather not have done or experienced, and our brains allow us to suppress or downplay their importance, to preserve or ensure our happiness. Usually, that's fine; dwelling on our flaws and mistakes can damage our confidence and wellbeing if taken to excess. That's a key feature of clinical depression.[1] On the other hand, persistently ignoring or skimming over bad, unhelpful or unflattering information can eventually become misleading, even dishonest. And, at this point in my investigation, I'd started to worry that I was guilty of this myself.

Basically, there are many things I'd ended up not including in the previous chapters. Charlotte Church being hounded

---

* World Wrestling Federation, as it was known then, during the era of Hulk Hogan and Macho Man Randy Savage.

by the UK tabloids when they arbitrarily decided to turn on her; Girl on the Net's dealings with men who get homicidally outraged when told that women don't 'owe' them sex; Lucy Blatter's tales of the ludicrously petty rivalries between the New York elites; Professor Chambers' mentions of the same thing in the world of neuroscience; Ian Boldsworth's increasingly common interactions with rude, abusive and intolerant audience members, and so on. In my defence, if I were to include *everything* I'd found or been told about happiness and the brain, this book would make the *Game of Thrones* series look like a pamphlet. Obviously, some things had to go. And I'm writing a book about happiness here, so didn't want it full of negativity or unpleasantness; that wasn't the narrative I was going for, so I left out the grim stuff where I could. However, it eventually dawned on me that the narrative of the book so far risked portraying humanity as a hard-working, safety-loving, mildly hedonistic wannabe-romantic bunch who just need to be liked and accepted, no matter what.

That's not true though, is it? Humans are often downright awful, sometimes because they're made happy by doing or experiencing things that are unpleasant, dangerous, or just plain nasty. What's going on there? Why would the brain cause us to feel pleasure and reward from unpleasant things? I begrudgingly realised that if I wanted to achieve a thorough and robust understanding of how happiness works in the brain, I was going to have to try and figure out the answer to that. I was going to have to go full Anakin Skywalker, and embrace the dark side.

## You say potato

The definition of unpleasant is 'causing discomfort, unhappiness, or revulsion', so logically you cannot be happy when experiencing something unpleasant. So, why do so many seem to enjoy such things regardless? In many cases, the answer to this is easy; that's not actually what's happening. What's bad or unpleasant is often highly subjective. An obvious example (one that I use a lot but it is undeniably useful) is food and food preference; the mere thought of a certain food may cause you to gag, while many others can't get enough of it. Things like oysters, or blue cheese, or tongue, or marzipan, or whatever; some foods seem to straddle the line between yuck and yum, and which side they come down on is literally a matter of personal taste. It's unsurprising when you consider how variable taste perception and preference is,[2] not just from person to person, but from situation to situation *for the same person*. Air pressure affects taste (part of the reason that airline food is the perennial butt of jokes), pregnancy and all the associated hormonal and chemical changes wreak havoc with it, age affects it, even simultaneously smelling or seeing something else alters how food tastes. The first bite really is with the eye.

Taste is actually a pretty feeble sense. The brain doesn't devote a lot of resources to it, so when we experience how something tastes, much of it is coloured by smell, vision, memory and expectation. Food preferences are therefore heavily influenced by experiences, preconceptions, culture, and so on.[3] So, when someone else eats something you think is bad, it's usually because their perception of it differs from yours. They *don't* hate it; it isn't bad to them.

This applies to the other senses too. Some people can't abide the smell of pipe smoke, while for others it reminds them of a beloved grandfather, so evokes only fond memories and positive associations (especially as smell is closely linked to memory[4]). Some can't stand heavy metal music, others live for little else. People regularly mock the fashions of the seventies, but perms and flares were hugely popular at the time. Basically, you can't just point at something you don't like and say it's definitively bad; that may be the case for you, because your brain has formed in such a way that means you find that thing loathsome. But other people have different brains. They *aren't you*.

It can't be emphasised enough how much variation there is between two individual brains. Because of this, neuroscientific (and related) studies regularly use identical twins as subjects,[5] because they have practically identical genes and grew up in the same environment, so were subject to the same conditions during their developmental stages. The nature *and* nurture variables are roughly the same for both. So if, as adults, one twin ends up with, say, depression, but the other doesn't, you can look at how they differ and more reliably conclude that it was this difference that led to the depression, because if it was a genetic or developmental thing then they should both have it. Or, if *both* end up with depression despite living different later lives, a genetic or developmental factor is likely to be the overall cause.[6] It's more complex than this brief summary, of course, but still, identical twins are a great boon for science, not just horror films.

However, even identical twins can be very different people, with markedly different brains and personalities. How? Think of it this way: get a million dice, pour them into an industrial washing machine, and spin for twenty minutes (while wearing earplugs, as that's bound to be noisy). When done, pour all the dice onto the floor, then calculate the total of each number shown. Then, do it all again, exactly as before, and work out the second total. Think you'll get the exact same total twice? You won't. Same dice, same machine, same procedure for the same amount of time. Nonetheless, it would be miraculous to have the exact same outcome twice. That's because, despite the overarching similarities, the individual components are all affected by random chance movements, and are constantly affecting each other in turn. Our genes and environment producing our eventual brain is a bit like this. Except there are a thousand billion dice, they all have a thousand sides, and the washing machine is on a roller coaster.

No wonder you get substantial differences from person to person. We've seen how people prefer different homes and living spaces, have different wants and desires when it comes to their career or ambitions, laugh at different things, have wildly varying sexual preferences and physical attractions, and so on. Nobody is 'wrong' here, nobody is doing anything 'bad', it's just that no two people are exactly alike, and what makes them happy will vary accordingly.

However, some influences are persistent and enduring, effectively 'loading the dice' in favour of an outcome; you grow up in a very musical family, you'll be surrounded by music all the time, so will probably have strong feelings about music. You might love it, you might rebel and hate it, but you'll probably not be ambivalent about it. Other influences, while temporary, can be incredibly powerful and engage many areas of the brain in significant ways, such as your first sexual encounter. Someone whose first sexual experience is with a redhead may well end up persistently attracted to redheads. The brain is quick

to learn novel things with highly stimulating, emotional properties,[7] so in this instance the basic learning processes rapidly
make a 'redheads = sexual pleasure' association. The brain is
good at generalising here; it doesn't have to be the exact same
redhead each time, because similar stimuli can produce a similar (if reduced) reaction,[8] resulting in an overall fondness for
things that have preferred elements in common. That's why we
like certain bands, or styles of music, or genres of art or film,
rather than just one specific example that we first discovered
and enjoyed. This does mean that if someone likes a thing you
hate, then there's a greater chance of them liking other things
that you are more likely to dislike. Differences between you
become wider and more ingrained.

However, before we wander down the 'everyone is different and that's cool, peace and love to all' path again, there
are indeed many things we can do that are *objectively* bad, as
in harmful to us, that countless people still enjoy and derive
happiness from. Given how risk-averse and safety-obsessed
our brains supposedly are, why do people enjoy unhealthy
foods, alcohol, drugs, gambling, dangerous violent sports,
and so on, despite being warned about them constantly? The
evils and dangers of drug use are hammered home from a
very young age,[9] the health risks of smoking are part of the
packaging,[10] same goes for the chemical and calorific properties of our food. The next superfood-rich colon-cleansing
immune-system-boosting diet is seldom more than ten
minutes away, waiting to make you out to be some form of
puppy-killing monster if you so much as look at a pack of bis-
cuits. And yet, we persist. Why?

Once again, it's because the brain doesn't do things 100
per cent rationally. For instance, although we're constantly

made aware of how unhealthy or dangerous certain things are, 'awareness' isn't all that helpful. Social media is often awash with some story, meme or game being shared to 'raise awareness' of a health condition or tragic event. Even if we accept it as 100 per cent well intentioned, many have pointed out[11] that once you've raised awareness, then what? Abstract awareness of something, even if it's a hazardous something, rarely changes actions or behaviours. This is a big problem for those trying to tackle health concerns like obesity, or major environmental issues like climate change. Even if people know something is wrong or harmful, they seem to persist in doing it regardless.[12]

This is partly because our brains, as powerful as they are, still have limits. Modern life means we're bombarded by information of all sorts at every waking moment, but the brain can only deal with so much information at any given time. The fact it manages to absorb and retain all that it does is borderline miraculous already, but it means the brain must pick and choose what's important, and ignore, downplay or just disregard the rest. How does it decide what to focus on?

A lot of the time, information with a significant emotional element* or stimulating properties (leading to 'arousal'[13]) takes precedence over more neutral information that lacks these qualities. So, if we eat deep-fried cheese nuggets or a triple-decker chocolate pudding, it tastes *gooooooooood*; we experience pleasure and enjoyment, because our brains react positively to sweetness and/or high-calorie food.[14] Therefore, our brain quickly learns deep-fried cheese = good. Contrast that with being told, via some pamphlet or dry documentary,

---

* Scientifically termed 'emotional valence', which can be positive, like for joy, or negative, like for fear.

about the long-term effects of fatty foods on our cholesterol levels and arteries. Potentially interesting, but nowhere near as stimulating, as *arousing*, as actually eating the stuff. So we're aware that eating fried cheese is 'bad' in some abstract way, but we *know* it's extremely pleasurable. And the latter has a better chance of influencing behaviour.

This also explains why, unless you're passionate about such things (and many are, admittedly), learning about science or maths or anything like that is difficult; it's mostly abstract, intangible information (by necessity), with little or no emotional or stimulating elements. We can still consciously work to retain it via repetition and revision, but this takes effort and persistence. It's a lot of work for no immediate, tangible reward, which makes it even harder again, because the parts of our brain that monitor such things don't approve of this. That's why I can recite my favourite *Simpsons* episodes word for word many years later, but I've no memory of what was covered in, say, my last school geography exam. Only one of those was important for my academic success, but the relevant parts of my brain clearly didn't like it. We haven't evolved to work that way. And, once we've decided we like something, we're reluctant to change our minds unless the counterargument is particularly strong.[15]

It *can* be done, of course. You can love cars and driving, but experience a near-fatal crash and it could be a long time before you set foot in one again.[16] Similarly, if we eat something we've always liked and get food poisoning, it'll be a while, if ever, before we eat it again. We still have those parts of the brain that recognise and emphasise disgust and danger, which fire up when we do something self-harming. But they too are limited.

A lot of it is about timing; you put your hand on a hot stove, the pain is immediate and you reflexively recoil, having rapidly been made fully aware that the thing you just touched is dangerous and to be avoided. However, what if, due to some bizarre condition which meant the relevant nerves conducted signals at a snail's pace, you experienced the pain a week later? You wouldn't automatically associate the pain with the stove, so there would have been nothing stopping you from repeatedly touching it in the interim. Anyone watching you would think you were bonkers and self-destructive, but you wouldn't know any better.

The longer the delay between action and consequence, the more difficult it is for our subconscious learning systems to make the connection.[17] Sadly, if we eat fatty foods or overindulge in alcohol or other drugs, the negative consequences like ill health occur days, months or even years after the event. Hangovers occur the next day, but that's long after the pleasurable effects of drinking have been experienced. The calorie-induced clogging of our arteries and pressure on the heart is very gradual, and we mostly can't even feel it. Point is, we 'know' it's not doing us any good, but the more primal yet still powerful brain regions that are more concerned with cause and effect don't really appreciate that.

In truth, even the conscious processes, handled by our frontal lobes, can be unreliable here, thanks to things like the optimism bias,[18] where we tend to assume a best-case scenario is the more likely outcome, based on nothing more than baseless assumptions. In many ways, this is actually helpful; a positive, optimistic outlook is reliably linked to improved mental wellbeing and tolerance of stressful events,[19] and can help with motivation and goals. On the other hand, assuming things will turn out fine can be unhelpful, even self-defeating. 'I could avoid getting lung cancer by quitting smoking, but I'll probably not get it anyway, so why bother?' – and then you get lung cancer. Because you smoke. See how that works?

This isn't just wilful ignorance on our part. Neuroimaging studies suggest that certain brain regions, specifically the amygdala and rostral regions of the anterior cingulate cortex,* seem to be highly active when subjects imagine positive future events, but not when they imagine negative future ones,[20] suggesting the brain automatically assigns more weight and importance to optimistic predictions than pessimistic ones. There's a certain logic to this; forward planning and predictions are, in evolutionary terms, relatively new things for our brains, and the deeper brain regions like the amygdala are just reacting to the basic qualities of what they're presented with, so emphasise the good things over the bad thing, without realising they're theoretical scenarios, not actual events. As a result, our predictions are often infused with unrealistic optimism.

There are processes, though, in which the brain tries to *prevent* us from doing ourselves harm. Studies on drug addicts and their long-term behaviours have shown that addictive drugs stimulate the dopamine reward pathway, that source of all the brain's pleasure and enjoyment. Over time, this activity diminishes; the ever-plastic brain changes to compensate for the constant presence of the drug, and so it takes increasing doses to induce the same highs as before, because the responsiveness of the reward pathway to the now familiar outside chemical is reduced.[21] It was previously assumed that

---

* Areas we've mentioned previously due to their prominent roles in aspects of emotion and reward.

this diminished reward activity causes drug users to maintain their habit, and that altered links between the reward pathway and frontal cortex regions responsible for consciousness, thought and behaviour mean addicts end up prioritising the monkey on their back over all the more 'usual' needs, like sociability, food, hygiene, and so on.[22]

Now, though, recent studies have pointed to the existence of an *anti*-reward pathway, a network of brain regions that causes negative emotional and physical reactions to things, even things we enjoy.[23] It's less well understood than the reward pathway, but seemingly involves certain regions of the amygdala and the stria terminalus (near the thalamus), has links to the frontal cortex, and relies on the neurotransmitters corticotropin-releasing factor (CRF) and dynorphin.[24] CRF has been found to be abnormally high in the spinal fluid of people who have died by suicide,[25] and dynorphin has been repeatedly linked to stress and depression.[26] Both are believed to cause dysphoria, a profound state of unease and dissatisfaction, essentially the opposite of euphoria. Basically, this anti-reward system makes us unhappy.

Weirdly, it's apparently activated when we experience pleasurable things, albeit much less than the reward pathway (at first). We experience intense pleasure at something, but also a hint of displeasure, as the brain effectively 'reins us in'.* However, studies suggest that chronic drug use gradually increases the activity of the anti-reward system, while reward system activity diminishes. Too much drug use can

---

* Or, it may be just to keep the opposing system functional. A lot of biological functions are controlled by two opposing systems, like the sympathetic and parasympathetic nervous systems, and there needs to be a baseline level of activity on each side to keep the component cells alive.

throw out the delicate equilibrium, so addicts end up with a barely responsive reward system and a seriously overactive anti-reward system. Eventually, drug addicts find it very difficult to be happy, but they can be seriously *un*happy. Their brains have been thrown out of whack. That's why long-term users don't persist with their drug use for pleasure; many of them, by their own admission, just want to feel normal again, and their drug is now the only thing that quietens the anti-reward system in their now-altered brain.[27]

This also explains why stress-induced relapse is so common in addicts; the anti-reward system works largely via the stress response mechanisms,[28] so stressful events increase the anti-reward system activity even more. Assuming all brains have this anti-reward system (and we've no reason to think otherwise), and everyone's life involves stressful things to some extent, then that would be another reason why people indulge in harmful but pleasurable things; it's not hedonism or indulgence, it's a genuine, if maybe subconscious, effort to stop being unhappy. Drinking, smoking or unhealthy foods are bad because they can cause you harm, and so make you unhappy. But, if you're unhappy anyway, what have you got to lose?

## Do unto others

So yeah, people regularly do things that damage their bodies and brains. But, in fairness, what could be more 'theirs' to do with as they please? If they're not harming anyone else, what's the problem? Problem is, they often *are* harming someone else! Passive smoking, drunken aggression, self-induced ill health causing a needless drain on precious medical resources that are there for everyone, etc. And that's just the incidental stuff; every day, people actively lie, cheat, attack, steal, bully, manipulate and sabotage, just to get what they want. Their goals and desires, their happiness, involve making people unhappy, often considerably so. Doesn't this fly in the face of what we've concluded before now? We've seen how being liked and accepted by others is a major factor in what makes people happy, and that even minor pleasant social encounters trigger the reward system, while even slight rejection causes us (psychological) pain. Then there's empathy, which, while being very useful overall, means we can experience other people's unhappiness ourselves, albeit to a reduced degree. Making others unhappy would logically make us unhappy too, wouldn't it?

We've even evolved distinct emotions – shame and guilt – specifically to make us feel bad about hurting others. While often used interchangeably, they are two separate things. Shame is directed inwards; it focuses on the self, producing a feeling of regret and unhappiness because you are aware that you have failed to live up to your own expectations and standards. Guilt, by contrast, is more external; it's caused by an awareness that others are being harmed in some way by our actions. Both are supported by a wide neural network encompassing the frontal, temporal and limbic areas of the brain.[29] In the temporal lobe, shame produces activity in the anterior cingulate cortex and parahippocampal gyrus, while guilt is more linked to the fusiform gyrus and middle temporal gyrus. Shame also produces activity in the medial and inferior frontal gyrus in the frontal lobe, where much of our sense of self and identity and self-assessment occurs. Guilt, by contrast, sparks activity in the amygdala and insula, where more 'external' issues and dangers are recognised.

All very interesting, but the point is that we have these many complex and entrenched neurological mechanisms to compel us to be nice and treat other people fairly. Yet a lot of the time we overrule or ignore them, and cause harm and hurt to others on levels that range from trivial to brutal, all in the name of getting what we want. What's going on?

Sometimes, 'tough love' and 'you've got to be cruel to be kind' are valid responses, and you have to cause harm to someone in order to help them in the longer term. Slicing someone open and rummaging around with their innards usually isn't anyone's idea of a friendly gesture, but surgeons do it every day, all in the name of saving lives. And there's the subjective element again; what some people consider antisocial or hostile behaviour is intended as anything but. I once spoke to an evangelical Christian who used to stand in the street on weekends and preach to the crowds of shoppers, telling them to accept Jesus and repent their ways. Haranguing innocent pedestrians with claims about hellfire and judgement when all they want to do is buy some new shoes – what's friendly or 'Christian' about that?

A lot, actually. These Christians genuinely do believe that only by worshipping God wholeheartedly can you get into heaven when you die, so anyone who doesn't is going to spend eternity in hell. Therefore, the decent thing to do is prevent this, by convincing people to join your church and belief system. It's the theological equivalent of ushering people to the lifeboats if you know the ship is sinking, even if it means spoiling their cruise. You may not agree with those preaching in the street, but from their perspective they are doing you a favour. They are doing good, according to the guy I spoke to anyway.

And yet, despite these caveats and all the brain's defences against it, there are still plenty of times where people do things they know will negatively impact on others, but that will lead to their own personal gain. In these situations, how do we end up listening to the devil on our shoulder, rather than the angel?

That metaphor is a useful one, actually, because a lot of the time (like with the reward and anti-reward system arrangement just described) there are different parts of the brain working to produce opposite outcomes, and which one ends up dominating varies from situation to situation. So we have these brain regions compelling us to be nice and friendly, but also different areas encouraging an every-man/woman-for-themselves approach. For instance, a 2011 neuroimaging study by Luke Chang and colleagues[30] revealed that when playing a game involving receiving money then deciding how much to return, subjects who gave back the expected or requested amount showed raised brain activity in areas linked to guilt processing, like the insula, while those who kept more money than requested showed raised activity in regions linked to reward, like the nucleus accumbens. Among the many useful insights from the study, it provides evidence that *anticipation* of guilt is a potent motivator for behaviour; the mere possibility of guilt was enough to compel people to return the full sums. However, some people are less sensitive to guilt than others. If the possibility of reward is more stimulating than the possibility of guilt, then you would put your own needs and desires before the wellbeing of others more often. And probably end up becoming quite rich in the process. Obviously, this would lead to a world where the richest people are often cruel and largely self-absorbed. Can you imagine!

It's also worth noting that these neurological mechanisms that compel us to be nice and friendly are relatively new, in evolutionary terms. Those concerned with self-preservation and gratification are older, more 'established', because deep in our evolutionary past we *were* simple, primitive creatures trying to survive in a dog-eat-dog world. The benefits and rewards of being part of a big friendly social group came later, when our brains were already developed to a decent extent. This becomes clearer when you look at things like the orbito-frontal cortex – remember that higher-reasoning region that pours cold water on our basic lustful impulses in situations where they would cause trouble and upset further down the road? A bit like the more complex brain regions yanking the leash of the more animalistic parts and shouting, 'Down, boy!'

Another example is the supramarginal gyrus's role in empathy. Our brains are egocentric; everything we do or think is experienced from our own perspective, so a lot of the time we view others and their actions through the filter of what *we* would do, or think.[31] While understandable, this can be unhelpful when dealing with others, because, you know, they're not us. This is especially true when it comes to empathy, working out what other people are thinking or feeling, because our own feelings can cloud things and confuse matters. However, a 2013 study conducted at the Max Planck institute by Georgia Silani and colleagues revealed that the supramarginal gyrus, another region located at the junction of the parietal, temporal and frontal lobes, essentially 'unscrambles' the egocentric distortion when it comes to empathy.[32] Think of it like the brain putting on 3D glasses; the jumbled, chaotic image on the screen is now rendered clear and decipherable, because the individual eyes are now receiving the images that make sense to them. The supramarginal gyrus is the 3D glasses of the brain's empathy system. However, it can only do this to an extent; if our own emotional state is wildly different from that of the person we're observing, the supamarginal gyrus has a lot more work to do, so we become a lot less accurate with regards to inferring the other person's emotional state.

Why's this relevant? Because we're less likely to care about upsetting someone if we *can't tell that they're upset*. So, if we're really happy, it's harder to recognise that someone else is unhappy, *even if it's us that's making them unhappy*. Quotes like 'Oh, he doesn't mind really' or 'Why can't she take a joke?' being applied to annoyed or enraged victims of our self-serving actions are common. It also explains why people having a fun night out get resentful of homeless people asking for change (something I've seen a lot). Their enjoyment means they can't fully grasp just how desperate and miserable the person must be to approach strangers and beg for money, so they perceive them as an annoyance, reacting with hostility rather than sympathy. Not fair, not nice, and not something we're powerless to resist (it's entirely possible to be considerate towards people far worse off than you), but it does suggest there's a neurological explanation for people acting, to put it mildly, like a bit of a dick.

There are, surprisingly, several ways in which our brains have evolved means of ensuring social harmony and happiness, which regularly backfire and cause the opposite. For instance, our brains seem wired for fairness. When we're treated fairly by others, it activates the reward pathways in our brain, much like eating chocolate or being paid money does,[33] and perception of unfairness causes significantly raised activity in the striatum, our old friend who's all about social acceptance and approval.[34] An evolved desire for and enjoyment of fairness would obviously be a huge advantage for any social creature. However, while this works fine when it's a small group sharing out berries or recently acquired meat, our societies are huge and complex now; we don't see what goes on in the tangled webs of infrastructure or behind closed doors, so we only have limited information to go on. As a result, we now regularly see unfairness where there isn't any. For example, it's very common for people to attack those who receive financial aid or support from the government. Those doing the condemning don't see the strife and terrible fortune experienced by those who desperately need help. No, they just see people getting free stuff. Free stuff they don't get. Stuff they are in fact *paying for* via taxes and all that. And that's not fair! That it's harder to empathise with someone worse off than you obviously won't help this persistent but inaccurate bias.

Speaking of biases, there's also the 'just world hypothesis', which describes the persistent belief that the world is not random or chaotic, but fair and just, that good deeds are rewarded and the bad punished. Given that the human brain has an innate liking for fairness and a tendency to expect the best, this belief in a just world makes sense. There's evidence implying that the insula and somatosensory cortex are responsible for this belief, at least to some extent.[35] Again, this implies that a belief in the fairness of the world may be innate to our brains. Like the optimism bias, this could potentially be helpful; assuming that good actions are rewarded and our efforts will be recognised would be a potent factor in motivating us towards long-term goals.

Problem is, the world *isn't* fair. Bad things happen to good people for no reason and awful people are regularly big winners in life. So, when confronted by these examples, it causes a dissonance; we believe the world is fair, but being confronted by an innocent victim of sexual assault, or seeing an utterly vile and immoral person become a multi-millionaire, clearly contradicts this belief. To reconcile this dissonance, we have two choices; completely change our belief system, calling into question the very nature of how we see the world and possibly overriding a bias baked into our brains, or we can work out why actually it *is* fair! And that's what we often instinctively do. That woman who was assaulted? She was asking for it! Dressed provocatively, she was. That evil millionaire? That's what business does to you, it's a harsh world, and he does provide jobs for many people, so a few assassinations and orphanage-burnings are fine! And so on.

There's also the common attribution bias,[36] meaning we blame the misfortune of others on their own incompetence or poor decisions, while if the same thing happens to us we attribute it to bad luck or circumstance. The more the other person has in common with us, the more potent this bias is. If they're the victim of a famine or volcano in a distant country, we've no trouble thinking of them as innocent victims, but if they're a lot like you, it becomes much harder to distance yourself from their misfortune, meaning it feels all too possible for you to suffer the same fate. One way to reduce the anxiety and fear this realisation causes is to assume they're just an idiot, and only have themselves to blame. That way, you don't need to worry about it happening to you, because you're not an idiot.

Our brains have all these properties and mechanisms to ensure we're as nice to others as possible, as well as remaining optimistic and motivated. Perhaps in more primitive times that was enough, and these things allowed us all to be happy. But in the modern world it's incredibly easy for events and factors to combine in such a way that an innate love of fairness and an optimistic outlook end up being counterproductive, and we end up harming others, often without meaning to.

All well and good. However, let's not overlook one important fact; people often *do* mean to harm others. Because they like it. It makes them happy. Why?

## I'm happier than you!

A stranger tried to fight me once. I was eighteen, had just started university, and was in a kebab shop with my new house-mates after a night in the pub. I happened to look across to the noisy group opposite, one of whom saw me looking, became enraged by my glance, and drunkenly challenged me to fist-icuffs outside, repeatedly saying it would 'make his night' to fight me. Thankfully, I was so confused I just stared at him, trying to work out what he was on about, which he apparently took for bold defiance, so backed down and returned to his chips. But from that night to now I often wonder, what was he getting from that? Why was the idea of inflicting physical violence upon a complete stranger so appealing to him?

Admittedly, I was wearing a bright orange shirt (can't remember why I thought this looked cool, but that's students for you). The surreal field of colour psychology, which argues that specific colours affect our mood and behaviour,[37] suggests that orange can induce low-level feelings of anger and hostility. Perhaps my wannabe aggressor was so drunk my

shirt was causing him distress? Wouldn't be the first time.*
Although have you noticed how violent prisoners are often
made to wear bright orange jumpsuits? Doesn't seem like the
best idea, in this context.

Colour psychology notwithstanding, there are times when
aggression towards a fellow human is valid. If they're attack-
ing you, or others, it's only natural to try and stop them by any
means necessary, and those means may well be violent. But
there are also times when we friendly, sociable, cooperative
humans who just want to be liked, opt to hurt or harm others
who don't deserve it. Our happiness sometimes *depends* on it.
A disturbing thing to think, but it's true.

Sometimes, it's a matter of logic. Our happiness can be
incompatible with that of others. Fair enough if being the
world's best gymnast will make you happy, but for you to
achieve this goal, everyone else who wants to be the world's
best gymnast must fail and be denied their dream. Similarly,
if having the most money, the top job, or winning the heart
of the most beautiful man or woman is what you need to be
happy, that means nobody else can have those things. There's
not enough to go around. Someone will have to lose out.

That, however, is where the brain does get involved. While
humans like to live in large groups and communities, these
groups, as with those of many social creatures, have a hierarchy. Sure, we want other people to like us, but we also want
them to admire us, to *look up* to us. Basically, we have an
instinctive need to be *better* than others. It's a deeply embedded instinctive drive, not some childish impulse.

---

* I also went through a phase where I thought bright, garish Hawaiian-style shirts
were cool and hilarious, so had amassed quite a collection, which weirdly disap-
peared around the time my wife and I moved in together.

Social hierarchy occurs in a very wide range of species, from mice to fish and beyond,[38] and drives a lot of behaviour. Dominance and subservience in groups is a huge part of life and community structure for many social creatures, from the alphas at the top to the outcasts and punching bags at the bottom. Why would humans be any different? If anything, our social hierarchy played a big role in making us what we are today; navigating a complex social structure may be what drove us to evolve such big brains. Understanding your place in the hierarchy requires self-awareness and an ability to grasp your position relative to others, and *raising* your status, meaning you get more rewards and, presumably, mating opportunities, requires guile, cunning and forethought. These are all complicated, difficult processes requiring a lot of brain power, especially as you're dealing with similarly smart individuals trying to do the exact same with you.

Studying how the human brain handles social hierarchies is difficult; getting a diverse group to interact while inside an fMRI scanner is a big ask, at the very least. But studies into primates like macaques reveal that changes in social status cause notable physical changes to brain regions like the amygdala, hypothalamus and brainstem.[38] These are deep, central, fundamental brain regions, so if human brains are even slightly similar in how they handle social status then it's clearly a very potent factor in our thinking and behaviour, affecting us at the deepest levels of our being. A related but separate network of regions in the temporal lobe and prefrontal cortex seems to activate when the more cognitive aspects of social status come into play,[39] suggesting a key role in planning and execution of our goals and behaviours.

And here's an important consideration: social interactions

may well prove rewarding, but there's evidence to suggest that social status modulates this, meaning that interactions where we increase our status over someone else are *more* rewarding,[40] and thus more enjoyable. Outwitting someone in a humorous exchange, getting promoted over your work colleagues, parents passive-aggressively pointing out that their offspring are doing better than those of others, getting more likes or retweets or followers than a rival, the proverbial 'keeping up with the Joneses', and on and on it goes. I'm not judging here, because I'm no different. Even now, while writing this book, I still occasionally stop to check how well my last book is selling compared to those written by fellow authors and friends/mortal enemies. If mine's doing better, it's satisfying, especially if they're a more established writer. Why? I gain nothing from this 'achievement', it affects neither of us, there's no award for it, and I genuinely feel immature and childish about the whole thing. But, it means, in some vague way, that I'm *better*. Because humans, while usually friendly, are also competitive. We're very sensitive of our social status, and really like raising it. Basically, winning is fun. Winning makes us happy, makes us feel good about ourselves. But, in order to win, someone has to lose. And that's not nice for them. Despite how often we tell our children that winning isn't important, there are parts of our brain that will have no truck with that sort of thinking.

Unfortunately, this pleasure in raising our social status can easily turn nasty. It means we tend to enjoy someone with a higher status being 'brought down a peg'. Charlotte Church told me about how the tabloid press suddenly turned on her, for no obvious reason. She was widely praised and lauded, literally described as an angel, and inserted into countless people's lives. But, once the novelty of all that wears off, the public can still be entertained by the fall of an idol. Someone who was presented as their superior suddenly being vilified and criticised provides a visceral thrill and pleasure, because we get to feel like we're better than someone 'higher up' the hierarchy. There's a global industry based around exploiting this phenomenon, from 'trashy' magazines to sordid reality TV shows, all dedicated to building people up and tearing them down. When you've got a brain that's very responsive to social status, we can gain a lot of satisfaction from the high and mighty losing theirs.

That's why mocking or criticising others can be enjoyable. That's why there are 'negative' types of humour, used to deride or humiliate;[41] there's a scientifically recognised difference between 'laughing at' and 'laughing with' someone. Some people even exploit this to become successful in the first place, like your 'shock jocks' or controversial pundits or other types of high-profile 'villain'. You can argue that these people say or do dubious, immoral or controversial things just for attention, but they *do* get the attention; their (often bizarrely high-profile) platform means those who agree with their questionable views feel validated and accepted. At the same time, those who disagree, perhaps initially drawn by their sense of fairness being violated by this reprehensible person being allowed to say these things, get to feel they're better, superior, to someone who's supposedly higher in status. A very satisfying feeling, even if you don't realise it's happening. There are undoubtedly numerous other factors to consider, but it's a viable neurological explanation of why we 'love to hate', and why we do it so readily.

So, if we're part of a group, we want to be accepted by that group, but we also want to have high status in that group. One way to do that is to be the best at something everyone agrees is good, and/or to be one who represents the consensus the most. Let's say you're part of a weight-loss group, something very common in this day and age. In many ways, the competitive edge can be helpful; the group formed around the idea that losing weight is important, so whoever loses the most weight is 'the best'. Organised weight-loss clubs give out awards for 'slimmer of the week' and reveal members' progress in front of the group,* presumably for this reason; losing weight often requires lifestyle changes that are hard to stick to, for all the reasons covered earlier, so any extra encouragement or motivation is a potentially useful tool.

Some can end up taking it too far though, wanting to be the one who best represents the group ethos, going the furthest to do so, pushing themselves harder to lose more weight than anyone else because that means they 'win'. Thing is, other group members won't just sit there and let them dominate, they'll want to prove they're also worthy of approval, so they try harder too, and maybe do better. And then the first person ups their game, and then others compete with them, and so on. Soon, what was the norm, like avoiding snacks and opting for salads over fried potato and losing a few pounds a week, is shifted to the extremes, and everyone is jogging on the spot for twelve hours every day and surviving on a diet composed purely of kale, carrot juice, and the occasional sniff of a picture of a steak.

This is group polarisation,[42] the weird phenomenon where members of a unified group end up thinking and behaving in ways that are way more extreme than they would if they

---

* Or so I'm told. I've never been to one myself; I don't need to. I am a bloated Adonis.

were alone. Far from balancing out or broadening people's individual stances, being part of a like-minded group ramps them up, thanks to our need for acceptance, approval, higher status, and so on. Remember, our position in our group is a big element of our sense of self.[43] And if we're a low-status member, we're more likely to feel rotten about ourselves.[44] For other examples of group polarisation, where individuals in well-defined communities end up becoming very extreme and radical in their views, see all modern politics.

That our group membership forms our identity is another important factor in why people are often unkind to others. Remember Dunbar's number, or Charlotte Church's observation that her friend's approval meant more than that of millions of strangers, or Ian Boldsworth's counterintuitive joy in people not liking his output; all these show that while our brains react positively to the approval of others, it's not necessarily *all* others. We may want many people to like us, but there are plenty of people out there that we actively dislike, even if we don't know them. Remember in Chapter One, where we saw that that lovely friendly molecule oxytocin can, in some circumstances, make you more racist?[45] Evidence suggests that oxytocin heightens emotional awareness and sensitivity, but nobody said the emotions had to be nice ones.

Essentially, humans love being part of a group. Our brains have evolved to accept and encourage this. Nothing can stop you if you're part of a group. Except, you know . . . other groups. Other groups are a potential threat to yours; they look and sound different, and believe different things. They are dangerous! Social psychologists define these as ingroups and outgroups. Your ingroup can be pretty much anything, from religious to political to familial to fandom, but a lot of the time it's cultural, and yes, racial. We are born and raised in a specific culture and among people who look like us, so we identify with them, get all our ideas of how the world works from them, and so end up wanting their approval and admiration. Someone from an outgroup, who we don't identify with, they're a threat, they're the enemy.

Studies have shown the amygdala, which is still best known for its processing of fear, is more active in people with strong racial bias when they view faces of different ethnicities,[46] and other studies even suggest it's harder to empathise with someone of a different ethnicity who's in visible pain.[47] Luckily, we can and regularly do suppress these negative impulses about people who don't look or act like us, and living among them, encountering them regularly, seems to expand our definition of 'ingroup' and reduce unpleasant prejudices.[48] But a depressingly large number of people can't, or won't, do this. Taken to its logical conclusion, we see people from other groups as *less than human*. If we don't recognise their individuality or autonomy, then we've no reason to care about their approval or empathise with them, so they're 'fair game' for persecution and attacks for our own benefit.

As a result, a reliable way for someone to achieve acceptance, high status and validation, and therefore happiness of some description, is to attack or hurt those who are not part of the ingroup. So, they pursue aggressive, harmful behaviours aimed at those whose only crime is being different. Full-on homicidal violence is obviously the worst exhibition of this, but it could be anything, like publicly (and fraudulently) condemning or harassing a political rival, refusing to provide services or fair treatment to those whose skin colour or sexual orientation doesn't match your own, or simply picking a drunken fight with someone from an unfamiliar group, just because he happens to be wearing a fetching orange shirt.

## Try not to think about being happy

Here's the thing; the point of me investigating all this was to find out why unpleasant or bad experiences and behaviours can still make us happy. But, most of the time, they *don't*; they do the opposite. They make us feel crap. Case in point; spending so long reading about all this was getting me down. I've a strong constitution when it comes to dealing with the bleak and morbid (experienced cadaver embalmer, remember), but the endless painstaking analysis of why humans treat each other like dirt for personal gain put a dent in even my sunny disposition.

Things came to a head when I was reading Girl on the Net's reports of men who cannot tolerate the idea that women won't have sex with them despite their 'best efforts', and who often become aggressive and violent when denied. I'd found numerous factors that could lead to such appalling attitudes and behaviours. For example, it is a depressing fact of society that machismo and male status are frequently measured by number of female sexual partners (even more depressingly referred to as 'conquests' – a telling term as people who are conquered seldom enjoy the experience). So, men who don't get sex are lower status, which is upsetting for them. Furthermore, the sexualised female form is omnipresent in our media and advertising, making it nigh-on impossible to ignore sex and the related impulses, particularly if the male arousal is as vision-based as evidence suggests.[49] Add to this the existence of easily accessed online pornography, primarily aimed at straight men, presenting women as passive recipients of sex for any passing man, coupled with endless mainstream examples of beautiful women ending up with less physically impressive men because they're a bit nice to them. All of this (and more) could lead to certain men ending up with a world-view in which obtaining sex is a key part of their identity, and a belief that women should, and will, provide it as and when required for any man who demonstrates the right behaviour, or says the correct chat-up line – like a hotel safe opening up when the correct numerical code is entered.

The thing is – and you'll have to bear with me on this, as some seem to find it a little complicated – women *don't* lack autonomy and individuality in this manner; they're humans (imagine that!), with inclinations and decision-making abilities of their own, and will hardly want to be intimate with any man who considers them subhuman, little more than an elaborate passion-receptacle. Accordingly, the expectations of such men are regularly thwarted; the 'effort' they've put in is not rewarded. The brain reacts very badly to all this, resulting in anger and hostility towards women, the finding of like-minded groups (usually online) to share their frustrations with, and then group polarisation kicks in and they end up hating women as much as possible and considering them the enemy and . . .

. . . And then I thought, 'To hell with this, I need some air,' and went for a walk to the nearby lake, in an effort to clear my head of the grim feelings I was currently having about my species.

As ever, as I was walking I was listening to one of Ian Boldsworth's podcasts, this time the aforementioned Parapod with fellow comic Barry Dodds, in which Ian and Barry discuss a particular ghost story or mystery or conspiracy theory, with Barry defending it as real or true and Ian invariably tearing him to shreds over it. As I listened to them bicker, Barry brought up his enthusiasm for ghost hunting, and his love of horror films and 'video nasties', like *Cannibal Holocaust*. And once again, this got me to thinking.

Many people actively enjoy things meant to scare and horrify, even though the whole point of these feelings is to deter people from whatever caused them. Even though the mighty human brain does have limits, strict logic is not one of them: it's easy for people to experience pleasure from things that scare them, to be compelled to seek out things we know are wrong and immoral, and to believe in things that have no rational basis. That's not abnormal for many humans; that's Thursday. How many books about serial killers are there? How often do people put themselves in harm's way purely for the thrill of it? There's no obvious social element here, and the fear is immediate and visceral, so we can't just blame it on the brain being tardy with regards to making the right associations. So what's going on?

Rather than dive into the bleak literature again, I figured I'd just ask Barry Dodds directly about what it is in the grim and gruesome that he enjoys.

Describing someone as a big fan of gore, horror and the supernatural may conjure up images of someone with sunken, reddened eyes, a sallow complexion, twitchy demeanour, maybe unkempt hair, poor personal hygiene. Barry Dodds is nothing like this; he's a friendly, fresh-faced Geordie with close-cropped hair and an upbeat if long-suffering demeanour, most likely due to the constant mockery he endures for his interests and beliefs. He also had to briefly stop our interview to rescue his cat Sox, who'd got stuck in a nearby box. A sinister, creepy individual Barry is not. And yet, he's passionate about gore and horror and ghosts. Why?

'I've always been attracted to things that scare me,' was Barry's simple yet revealing explanation. His earliest memory of experiencing the thrill of fear was around age seven, while staying with his Nanna in Amble, Northumberland, on the seafront.

'My older cousin Sarah stayed with us too, and she would tell me ghost stories about the nearby pier and promenade, about how there was the spirit of a monk who haunted there, and it used to *terrify* me.' During a later visit, when he was around thirteen, Sarah also showed him his first horror film (*Hellraiser II*, an unsettling sadomasochistic eighties gorefest). He was so scared he bailed on it half way through, but watched it the next day, and then every day for the rest of the visit, because as Barry puts it, the thrill, the exhilaration, caused by the fear was compelling.

It may sound like a child being psychologically bullied by an older cousin, and maybe it was, but what Barry's describing here is excitation-transfer theory.[50] Intense stimulation, particularly that caused by fear and the associated fight-or-flight response and all the adrenaline in your system, causes a heightened arousal and sensitivity to stimulation that lasts beyond the source of the original fear (it takes your system a while to return to normal). As a result, previously neutral things become more stimulating, because your brain has been 'knocked up a gear' and everything becomes more vivid. The excitation of the scary thing is transferred to other things that would usually be mundane. Hence the name.

Adding to this is the fact that our brain's reward pathways aren't just activated when something nice happens, but when something bad ('aversive', in the literature) *stops* happening.[51] Your subconscious brain essentially says, 'I didn't like whatever that was, but it's stopped now, so well done for avoiding it. Here, have some pleasure.' Our glad-to-be-alive brain is experiencing rewarding sensations that are heightened by the residual excitation. If you've ever seen someone all giddy and trembling after watching a horror film at the cinema, that's probably what's going on there.

There still needs to be an element of safety though; you need to know, on some level, that the danger isn't real, otherwise it will just terrify you, because it should. Very few survivors 'get a taste' for earthquakes or house fires. Barry's experience was of being terrified, but in a safe environment amongst trusted family. He retained a sense of control[52] making it more likely he'd experience the fun elements of being scared without the sense of danger, and ended up hooked.

Barry is also an enthusiastic ghost hunter, who does believe in ghosts, but doesn't believe he's ever seen one, so he spends many a weekend in purportedly haunted homes and castles with the surprisingly diverse range of tech and appliances available to modern spectre-seekers. But even this passion seems rooted in his enjoyment of fear.

'I don't actually know what I'd do if I *did* see a ghost. I'm guessing it would utterly terrify me, to genuinely see the spirit of a dead person, and what that would mean. I think it would scare the life out of me. But again, the fun part is the thrill, of being in the dead of night in an empty castle, and your heading towards some door that leads into the dark, and your blood's pumping and the adrenaline's flowing and your

hair is standing on end, that exhilaration . . . it's addictive.'
Barry's not alone in this; thrill-seekers are a recognised personality type for many scientists,[53] and some genetic evidence
suggests they have less responsive reward pathways than the
average person, meaning they may *need* the intense thrill of
dicing with death to experience the same pleasure as you or
I would obtain from, say, a decent cup of coffee, or a particularly well-made sandwich.[54]

But, where does the gore come into this? Enjoying being
scared is one thing, but that can happen without witnessing
blunt objects being forcibly shoved into human bodies in various ghastly ways, yet people derive a worrying pleasure from
that anyway. Surely such a thing would repel us? It does for
many people; haemophobia is a very real problem for some.[55]
But there are still enough people who enjoy it to make 'torture porn' a profitable cinematic genre.

There are explanations for this. It could be the same process experienced during horror and fear, that something
unpleasant has stopped happening so our brain recognises
this as a positive. Some say it's due to release of psychological
tension,[56] the perception then removal of brutal images causing distress, then relief, much like humour does, except in a
more grizzly, bloody way. It could be a novelty thing; we never
get to see such things normally, and witnessing them provides
a thrill. Maybe it's catharsis? Or an underlying curiosity which
means we want to see harmful things so as to avoid them ourselves later?[57]

All these things likely play a role, depending on the person. But something Barry pointed out struck a chord; he
has obsessive compulsive disorder, OCD. While he doesn't
feel this has anything to do with his interests in horror and

gore, there is evidence to suggest that those with OCD are more susceptible to dwelling on what some label 'forbidden thoughts'.[58]

If you've ever thought about cheating on your partner, or hitting someone who's annoying you, or pushing a friend off a cliff that you're both stood at the top of, or stealing some unattended money, then you've had a forbidden thought. They're the thoughts and impulses you have but really feel you shouldn't, because you know they're wrong or bad. But you have them anyway. Luckily, it's not because we're warped or evil; it's perfectly normal.[59] Remember that the powerful human brain can predict and imagine and anticipate events and outcomes by forming a constantly updating simulation, a mental model, of the world.[60] But the brain doesn't just sit and wait for things to happen; it's constantly testing limits and assessing options, much like the background processes whirring away on your laptop while you're typing. This means many possible options for action are considered in every situation, even hypothetical ones, and a lot of these options are going to be unpleasant, or wrong.

We have moral boundaries, ethical limitations and taboos, some of which are instinctive (like not wanting to be rejected by our group) but many of which result from our culture and upbringing. If you're raised in a strict Jewish family, idly thinking of eating pork would be a forbidden thought, but if you're an agnostic it's perfectly fine. In most cultures, inflicting serious harm on others is considered seriously immoral. And yet, it's *an option*, and thanks to all our baser instincts and compulsions, it's something we can, and do, think about. Often, these alarming thoughts appear and disappear rapidly, being dismissed as soon as they're produced. This is technically healthy, as it reinforces the limits of our mental model of the world; it's like the brain approaching a wire fence, hearing it hum, realising it's electric, and backing off. 'Can we go down this route? Nope! OK, let's try something else.' Forbidden thoughts may be the brain's way of checking where the boundaries are.

Problems can arise though, when people dwell on these thoughts, and give them more weight than is warranted. Those with an external locus of control, who don't believe they have much control over their own lives, seem more susceptible to persistent forbidden thoughts, likely due to a lack of self-confidence or self-belief. As mentioned earlier, people with OCD also seem particularly prone to this, dwelling on a thought that should be fleeting. That's OCD in a nutshell, really. But this leads to an unfortunate paradox; the more effort the brain invests in suppressing a thought, the harder it becomes to do so.

A 1987 experiment by Daniel Wegner simply asked subjects not to think about a white bear.[61] Those asked not to thought about it *way more* than those not given that instruction. These paradoxical effects of thought suppression are common. Have you ever tried to force yourself to relax so you can sleep? Or ended up thinking about eating much more than usual while on a diet?[62] What happens is, by trying to suppress a thought you don't want, the brain then turns it from a passive to a more active process, so more of the brain is engaged by it, making you more aware of it, and so it gains priority over other thoughts, and you start doubting yourself and worry about your actions, which makes you *more* concerned, so you dwell on it more, and on and on and on. Sometimes it becomes an all-consuming compulsion, affecting health and wellbeing. Sometimes people end up acting on these thoughts, which is . . . bad.

So, if you end up spending a lot more time thinking about doing or seeing unpleasant things, one option is to actually *do* them, providing relief and catharsis, but in a safe, harmless manner. Witnessing brutal atrocities on the cinema screen, reading about serial killers in the book bought for your train journey, gunning down crowds of people in an immersive video game; as much as people fear what these things do to us, they can make us happy by providing a release for these darker thoughts, impulses and drives that our brains are constantly coughing up, but that society declares we shouldn't have. Sometimes the best way to confirm boundaries and satisfy curiosity is to touch the electric fence.

There are possible downsides, of course. Maybe people do risk becoming desensitised by constant exposure to violent images and activities, meaning they could potentially end up craving the real thing eventually. There's no conclusive evidence for this though.[63] Our brains are still very good at separating real from unreal, so even if they do become desensitised to blood and gore on screen, it doesn't mean they'll end up behaving any differently.

Indeed, Barry admitted to me that his interest in gory films now is more to do with technical appreciation of realistic special effects done well and the display of innovative ways to harm the human body. Not that I'm one to talk; I recall many days when I was bored out of my mind while handling dead bodies. What does that say about the state of my own brain? Barry still gets terrified by ghost hunting though, and any films with a more 'psychological' element to the horror. But he loves that really, because he embraces what scares him. He even says his initial terror of public speaking is what led to him becoming a comedian, which led to him doing a podcast, and now he's in this book. Funny how things turn out.

That's the thing; every person and their brain is different, often radically, so there are many ways in which we can be made happy. It's just unfortunate that it often turns out that those ways involve harming ourselves or others. But we're all capable of bad things, and it's only natural to think of doing them from time to time. It's how much weight we place on these impulses that determine the sort of person we are, and how others end up seeing us. Not dwelling on the bad thoughts is ideal, acknowledging them for what they are then overruling them is often necessary, but if they become persistent and occupy a lot of your headspace, one of the advantages of human society is that there are ways to indulge them without harming anyone, be it through gruesome films, video games, or something like that.

A little catharsis and indulging of the darker impulses every now and then is important for overall happiness, as long as nobody gets hurt. It's when they *do* get hurt that we have problems. Your own personal happiness may be important to you, but is it more important than the happiness, wellbeing, even lives of others? It would be very difficult to argue that it is, no matter who you are. Not that this stops people, sadly.

I'm afraid there's no easy answer here. That's just the brain for you. Sometimes we should indulge our less pleasant impulses, other times we definitely shouldn't. It depends on situation, context, company, and numerous other things. But if there's anything that should be taken from all of this, it's that these dark impulses and thoughts are *normal*, so spending every waking moment trying to suppress or avoid them entirely is almost certainly going to cause you a lot of stress and distraction. Because we can't control all of our thoughts so thoroughly, and sometimes it's best to go with the flow. Ironically, this means 'don't worry, be happy' is very bad advice when it comes to mood. Bobby McFerrin has a lot to answer for.

# 8

# Happiness Through the Ages

In mid-2017, I was offered a free stay for myself and my family at the lovely Bluestone National Park resort, in west Wales, in exchange for a favourable mention in print. There, that's what that last sentence was. I'd been told high-profile media types often get sent gifts by companies and businesses hoping for a positive write-up, but in five years of writing for the *Guardian* all I had ever received previously was a voucher for some yoghurt. A jokey article I wrote[1] was spotted by a dairy company's PR person who told me it was her role to 'follow all the yoghurt-based news' (hearing that this was someone's actual job was reward enough for me). So a free holiday certainly made for a nice change. But the main reason I accepted was because, at this point, I'd spent so long researching and writing this very book, that I'd barely spent any time with my family. I figured they deserved something nice, if just to dispel my now ever-present sense of guilt.

This led to my wife and I discussing what activities to do, although my decision always came down to the same thing: whatever the kids would enjoy. The happiness of my children, currently aged five and one, means more to me than literally anything. These days I willingly go to playgrounds, or family pools, or play with toy spaceships, or watch hours of *Peppa Pig*. Twentysomething me would have shuddered at the very idea of these things, before returning to writing his half-baked comedy, reading sci-fi novels, and binge-watching DVD box sets.

Then again, hedonistic eighteen-year-old me would have

scoffed at the way twentysomething me squandered his independence by staying cooped up indoors when there was a whole world on the doorstep, much of which featured booze. And in turn, that notion would have terrified childhood me, who would rather have spent time with his comic books, or in the pool, or playing with toy spaceships. Basically, my children brought me back to where I started.

This is an intriguing point. I've covered a great deal about what makes us and our brains happy. But, what does so at one point won't necessarily do the same a year, five years, or even ten minutes later. That's why those scaremongering headlines about some new technology that 'changes your brain' are so misleading; *everything* we experience, from eating an apple to going fishing, 'changes the brain' to an extent. It's a fact of life; a static, fixed brain is useless in an ever-changing environment. A static brain is dead.

But doesn't this undermine the very concept of 'lasting happiness', or 'happily ever after'; things we're often told are the point of existence? How can happiness be permanent if the brain that creates it is not? What it comes down to is how extensive are these changes? How 'deep' do they go? Is it all surface level, like a television – ever-shifting images on a screen with permanent hardware beneath? Or, is it more like a caterpillar becoming a butterfly – a complete overhaul of everything, right down to the fundamental functions? The answer, presumably, is somewhere between these extremes.

And it's an answer I felt I should try and uncover, before I could draw a line under my investigation. So, as my last hurrah, I decided to look at how the brain differs and changes over the course of our lives, from birth to death, and what this means for our happiness.

## Happiness is childish

Technically, our brains never stop changing; every new memory formed requires a new connection to be made, and this process continues our whole lives. However, it's during childhood that our brains experience the most dramatic changes. What does this mean for our ability to experience happiness when our brains are in their most formative stages? Basically, what makes a baby happy? Because it's not like they can really *do* much except gurgle, sleep and fill nappies with borderline-toxic waste. Which is pretty weird, when you think about it.

Minutes-old horses can stand up unaided, albeit shakily. Tiny kittens or puppies, lacking vision or hearing, make their own way to their mothers to feed. And just-hatched turtles crawl across an entire beach, with their flippers, to get to the water, then navigate a whole ocean, alone. Compare this with human babies, who need help lifting their own heads up. If we humans are the smartest species, shouldn't we be more capable from the off? Why don't we exit the womb reciting Shakespeare, ordering lattes, and carrying a briefcase? Surprisingly, our big brains are to blame.

Basically, to accommodate our rapidly expanding brains, humans had to evolve bigger heads and skulls, which is why we *Homo sapiens* have much higher foreheads than our more-sloping-skulled cousins like Neanderthals.[2] But this growth was localised to our heads; our body size is consistent with primate averages. As a result, our physical development is essentially out of sync; our heads grow 'faster' than the rest of our bodies. A baby's body is around 5 per cent of its eventual adult size, but its head is around *25 per cent.*[3]

Because the dimensions of the birth canal are restricted by the width of the solid bone female pelvis, babies need to be born while their delicate heads still fit through it. But because evolution has caused our heads to develop at an increased rate, our bodies aren't as fully developed as they 'should' be when they emerge into the world. There are numerous theories about why human babies are born at the nine-month mark specifically, incorporating bipedalism, energy demands, even the invention of agriculture.[4] But whatever the cause, babies are born at a much earlier stage of physical development than most other species.

This could explain why so many people describe babies and their brains as a 'blank' slate, with no preconceptions or concepts. Technically, this isn't true; a newborn brain isn't an amorphous blob of brain cells, waiting to be sculpted by experience. Certain aspects of the brain are 'hard wired', like the functions of the brainstem, essential for life. Nobody needs to be taught how to breathe and excrete, thankfully. Evidence also suggests that a lot of sensory development occurs in the womb, even taste and smell. Babies are also born with reflexes like being startled, or the latching one for feeding, so obviously *some* brain development has already occurred.[5]

In terms of happiness, one important suite of neurological processes that develops very early, maybe even in utero, is that which governs emotional reactions. Babies cry as soon as they're born, suggesting an awareness of distress. They will stop crying when placed in their mother's arms, suggesting they are experiencing a sense of safety, maybe of comfort. Studies with orphaned chimps presented with inert 'replica' mothers showed that they tend to prefer ones covered in soft cloth rather than hard, unyielding ones, even if the latter are the only ones that feed them.[6] Primate and human babies and infants instinctively need contact and cuddles; it makes them happy, insofar as they understand such things. We've also seen that babies start smiling and laughing before they can talk and walk.

Evidence suggests the limbic system, that diffuse network of regions that meshes emotions, consciousness and basic instincts, forms very early.[7] This is particularly true for the amygdala, which we know plays numerous vital roles in our emotional processing, and studies show links between the amygdala and areas like the striatum and parts of the insula are present from the off and remain stable right through childhood and beyond. If we accept, as has been argued throughout the previous chapters, that the striatum is integral for much of our social cognition and awareness, and the insula is key for many emotional responses tied to sense of self, it would be reasonable to say that young child brains are able to experience the relevant emotional reaction to good and bad things, particularly in the context of other people. The tickling and peek-a-boo examples from earlier show just how young ones enjoy and appreciate interactions with a safe person. Babies and young children smile when they see a familiar person they consider benign,[8] but may cry when handed to a stranger they don't know, or don't like the look of. Exactly why they don't is anyone's guess. They are very little.

In truth, you could fill multiple books with discussions and observations and theories about how the human brain develops through childhood, and better scientists than I have done just that, but there are a few interesting neuroscientific and psychological aspects worth considering when it comes to our overall happiness.

An ability to recognise that something is bad or good, and to have that strongly reinforced via relevant emotional reactions, would be a crucial tool for learning about how the world works, particularly for a rapidly developing brain. Some estimates suggest the early years of childhood see the brain forming up to one million new neural connections *per second*! This results in rapid brain growth; a child's brain is half adult size at just nine months, three-quarters grown at age two, and 90 per cent adult size at age six.[9] A child's brain is acquiring new experiences, positive and negative, at a frightening rate. This helps explain why children are so inquisitive and curious about everything – whether it be your plug sockets, delicate ornaments and valuable devices, or the cupboards where you keep the toilet cleaner and paint thinner. We've seen how the human brain appreciates novelty, but to a very young child *everything* is novel! Every exploration and experience is forming new connections in their brain that may serve them for a lifetime to come. That's why they get into everything. It's also why they need to sleep so much, compared to adults; their brains need a lot more 'down time' to process everything they've acquired just by being awake.[10]

Even after the ferocious growth of the brain during these early years, you brain is never more pliant and absorptive than when you're a child. Because of this, many studies point to the danger of toxic stress.[11] The ability to experience emotions, including fear and distress, and to respond to social cues, forms in the brain almost right away, but the understanding and appreciation of context and situation is acquired far more gradually via learning and experience. As a result, children are very sensitive to stressful environments, like where parents row and shout, or scary events occur. They don't know the cause and what it means, they can't appreciate that mummy and daddy are just exhausted and arguing over whose turn it is to put the bins out; all they can grasp is that a bad, scary thing is happening and they can't do anything about it, something extremely stressful for any brain, let alone one so new. The subsequent wash of stress chemicals sent through the system can genuinely interfere with brain development and growth, leading to problems with cognitive development later in life.[12]

Luckily, this brain malleability can have positive outcomes too. One recent study[13] suggests that the environment you inhabit at around age four will significantly affect the structure of your brain at early adulthood. Specifically, the more enriching your environment when you're four, the more structurally developed your brain will be over a decade later. Why four is such an important age is hard to pin down, but it may be a key point in the brain's development. For instance, evidence suggests that our earliest memories begin around age four.[14] Maybe until that point the brain is still 'sorting itself out' in terms of important functions, so memory formation is less reliable? It's like readying the car for a long journey; you pack all your stuff in the boot, check you've locked the house, make sure the fuel tank is full, and so on. All important aspects of the trip, but you've not actually *gone anywhere* yet. Eventually you climb into the driving seat, shout 'let's roll!', and start off. This might be what the brain does, at age four. Or thereabouts. Metaphorically speaking.

However, there's still a long journey ahead and the brain still has a lot of developing to do. Theory of mind, that ability to grasp what others are feeling or thinking, seems to form quite early on, but becomes more elaborate and refined as children learn to laugh and empathise with others.[15] Childhood IQ also seems to be far more variable in response to environmental factors (different schools, teachers, peer groups, etc.) than adult IQ, which is more 'fixed'.[16] Children often *need*\* to be around other children they can interact with, and can be very susceptible to the effects of group membership – polarisation, cooperation, intergroup rivalries, etc. – but these can also be very easily reversed.[17] A child can have a blazing row with a friend over something trivial, with both vowing never to speak to each other again, only to have the whole thing forgotten the next day.

This tendency towards unpredictable or inconsistent behaviour is a common feature of childhood, as any parent trying to keep track of endlessly changing food preferences will attest. One potential explanation is that the links between the amygdala and the prefrontal cortex, where much of our rational thinking and higher reasoning seems to be based, seem to change drastically between childhood and adulthood. One extensive study[18] observed that a child's brain shows activity suggesting the amygdala stimulates the prefrontal cortex, implying emotional reactions could take precedence over logical thinking, which would certainly explain tantrums, or constantly asking the same question – 'Are we there yet? Are we there yet? Are we there yet?' – when they don't like the answer. You can say 'not yet' all you want, but if the child is bored and frustrated then that's what will be dominating their consciousness, not your logical answer.

However, later in adulthood life this connection essentially

---

\* Remember Chapter Four and the perils of social isolation? Or Chapter Six and the importance of play among young peers? Our brains seem predisposed to avoiding seeking these things out.

'flips', and now recorded activity suggests the prefrontal cortex can negatively affect the amygdala. Basically, our rational thinking can overrule our emotional responses; a vital skill when navigating the modern world as an individual.

While all this is interesting, much of the literature suggests the most important factor in a child's happiness is the relationship between child and primary caregiver. While obviously not always the case, this caregiver is most often the baby's biological mother; as well as being the one who created the baby inside their actual body, the baby–caregiver bond is strongly regulated by oxytocin,[19] which new mothers are awash with.[20] Indeed, some studies suggest that oxytocin, responsible for much of human interaction and happiness, originally evolved to encourage the bond between mother and child.[21]

This goes both ways, too; activity has been recorded in the brains of mothers* when watching their own child laugh or cry that is markedly different to that seen when watching other, similar babies do the same thing.[22] It seems their brains are very sensitive to their specific child and their emotional state. This parent–child bond runs deep.

It's also regularly the bedrock of a child's life, and the main factor that determines how their brain develops. To do all the exploring and investigating and interacting they need to do to learn how the world works and consequently be happy, children need somewhere safe they can retreat to if things go awry. Or, some*one* safe. Attachment theory is the psychological model that dominates much of the modern study of infant behaviour.[23] It states that infants will mentally 'attach'

---

* There haven't been many, if any, studies investigating this process in primary caregivers who aren't the biological mother, but that's not to say it doesn't happen in their brains too.

to the primary caregiver, and use them as the primary source of safety and feedback about how things work. How children respond when removed from then returned to the primary caregiver in a strange situation is an oft-used tool to assess the parent–child relationship and the functioning of the child.[24] The nature of this attachment is said to have far-reaching consequences, incorporating everything from personality type,[25] career development[26] even sexual orientation[27] later in life.

Psychologist Diana Baumrind attempted to define the ideal types of parenting back in 1971, and asserted that the best approach is a mix of permissiveness and discipline.[28] According to hers and subsequent findings, a child needs to be able to explore, to experience new things and make new friends, so allowing them to do so is important for their happiness. But, they also need to know where the limits are, to feel safe within them and to be able to learn that the world has rules. An important concept when it comes to pretty much everything.

It's sadly quite easy, at least in the neurodevelopmental sense, for parents to go too far. Too much discipline, pressure and punishment for 'wrong' behaviour can lead to children who may be high achievers, but who also think that approval and affection can only be obtained via performance and success, leading to high levels of neuroticism and poor social cognition, and even related disorders like bulimia.[29] Conversely, a too permissive and relaxed approach to parenting can lead to children having distorted social awareness. You may have seen those kids who are 'out of control', who are destructive and disruptive because their parents never tell them off. That's what you get. Such children often struggle to form meaningful relationships, because they don't follow social norms that other people expect and get rejected as a result. This obviously makes them unhappy. Similarly, a lack of parental reaction to behaviour can produce apathy and a lack of goals and ambitions. Your parent's actions and reactions are how you learn about the world, and if your parents don't react to anything you do, it's easy to see how things can end up seeming meaningless.[30]

Overall, there are many things that make a child happy, and a lot of these apply to adults. But because of the ever-changing nature of a child's brain, what causes happiness can be more fleeting or more intense, or both, and can shift rapidly from one day to the next. It's a chaotic existence in many ways, which is why the parent–child relationship is usually the core around which a working understanding of how the world works is built. It wouldn't be ridiculous, then, to argue that while there are vast numbers of other variables to consider, the parent–child relationship is possibly the most important facet of a child's happiness.

Ideally, the primary caregiver will be loving and encouraging, and consistent. Consistency is key because the child is deriving much of what it needs to know about the world and its workings from this caregiver. They may understand language eventually, but they learn just as much from observing and mimicking[31] and their powers of logic and reasons are still being formed and refined, so mixed messages delivered via words or behaviour are unhelpful. The command 'do as I say, not as I do' will just confuse a child, because they can recognise the hypocrisy.

This can be difficult, because life isn't consistent and parents are humans. Luckily, 100 per cent consistency isn't essential, it just needs to be enough so a child can get the overall point, and a good parent can explain and repair any deviation from usual behaviours caused by a period of fatigue or stress (both very common when you have children).[32] Basically, if you're nice to your children and set a decent example, odds are they'll be happy.

Of course, this is just a rough conclusion based on the available data I've seen. You may have experiences and information that differ completely. I'm not telling you how to raise your children here. I know people get really unhappy when you do that.

## Teenage kicks right through the brain

I was a teenage rebel.* Might sound an unlikely claim for a nerdy scientist, but it's true. What you've got to account for is what I was rebelling *against*. I did the usual adolescent thing of rejecting authority, but my authority figures were my parents. My father in particular was something of a tearaway in his youth. One time he accompanied me to my school parents' evening, and my maths teacher Mr Owen, who actually taught my father at my age, flat out informed my dad that there was no way I was his biological son, given the sort of things I did in maths class. Like turn up.

So, when I went through my teenage rebellion phase, that's what I was rebelling against. My dad would encourage me to go out and meet people, maybe chat up some ladies, and I was all 'Screw you, old man! I'm staying in to read a *book*!' This didn't have quite the same vibe as the stereotypical 'teenage rebel'. If anything, me wearing a leather jacket

* This was after I was a cheerleader.

just made it more embarrassing. Still, I guess it got me to where I am now.

In my defence, I was merely fulfilling a cultural cliché. Curfews, groundings, defiance, blazing rows, are all apparently common features of a typical parent–teenager relationship. But why? If a positive relationship with parents/caregivers is maybe *the* essential feature of childhood happiness, why the sudden and drastic change?

Adolescence is the transition period between child and adult. 'Adolescent' is usually taken to mean 'teenager', but the boundaries of adolescence aren't that clear. Part of adolescence is puberty, that hormone-induced process where we become sexually mature. However, puberty begins from around eleven to twelve years of age for boys, and ten to eleven for girls,[33] while evidence suggests physical growth and brain maturation continue into the mid-twenties, so while our teenage years are when most of it occurs, there's still debate around where adolescence begins and ends.

No matter. What's important here is the effect of adolescence on our happiness. And it's not great. While all the things that affect adult happiness should also apply to teenagers, they're often thought of as moody, grumpy, stroppy, angry, always listening to gloomy music, getting into risky and dangerous behaviours like underage drinking and sex, drugs, sleeping all hours, and so on. Basically, accepted wisdom is that teenagers *aren't happy*. Why? A lot of this is to do with the changes taking place in their brains.

Surprisingly, your teenage brain has *fewer* connections than your childhood one. This is because, while a child's brain may be forming millions of new connections every second, not all these will be useful. A child's brain is essentially hoarding; it won't throw anything away. However, while the brain can't get 'full', what all these superfluous neuronal connections do is hamper efficiency; the most capable human brains tend to be the most efficient, the most well connected.[34] A child's brain is anything but that. Which may explain why they're often so erratic and easily confused.

So, during adolescence our brains undergo a process called pruning.[35] It's pretty much what it sounds like; excess and unnecessary connections (synapses) and neurons are removed, deleted, while the ones you use regularly are retained and reinforced, improving the overall functioning of your brain. It can be quite a drastic process; estimates suggest that up to 50 per cent of existing neurons and connections are removed by pruning, no matter what the connections may have represented. For instance, when I spoke to the now-adult Charlotte Church, I asked about her memories of her childhood stardom, and she confessed there'd been times in later life where she'd been excited about meeting some big-name star, only to discover she'd *performed alongside them* in her youth! But, if you have a childhood so full of such incidents that they become normal and unremarkable, the pruning process isn't going to spare them.

This may seem wrong; how can a reduction in brain cells improve the brain? But that's like wondering how a classical statue can be considered superior to the block of marble it was hewn from; here, as with the brain, 'more' does not mean 'better'.

One potential consequence of this drastic brain overhaul is increased need for sleep; the average adult needs around eight hours sleep a night, but the typical teenager often needs nine, maybe ten.[36] Maybe teenagers would be happier if they just got enough sleep? Unfortunately, teenagers must still go to school, which starts early in the morning. Well-meaning parents often cajole fatigued teens to stick to a 'normal' schedule, haranguing them for sleeping the morning away. They may even be pressured to 'get a job', earn their keep. Even if they aren't, your teenage years are when exams happen, which determine *your entire future*, meaning those so inclined study all hours. Overall, adolescent brains genuinely need more sleep, but modern life means they seldom get it. Sleep deprivation is known to damage mood, happiness and cognitive functioning,[37] yet teenagers can end up dealing with this for years! Seems harsh to insist that they be happy about it.

Then there's puberty, with all the (often unpleasant) physical changes that causes: greasier skin and acne, unseemly hair sprouting in previously bald places, changing voices (for boys), the onset of menstruation (for girls), and so on. These changes are induced by the sudden dumping of sex hormones into our bloodstream.[38] But remember, sex hormones also influence sexual arousal, both via the sex organs and the relevant brain regions. So, you suddenly find yourself wanting sex, even though you're not exactly sure what that is yet, while your hormones are conspiring to make you look as weird, off-putting and awkward as possible.

Unless that was just me? Either way, this would clearly lead to more frustration than happiness.

Basically, adolescence means we need more sleep and more sex, but it also makes both harder to achieve. Surely that's reason enough for teenagers to be more hostile and less happy than children and adults? Perhaps, but it's not the whole story. Evidence suggests there's more going on, deeper in the brain.

An interesting paper by Professor B. J. Casey and colleagues from 2008[39] suggests a neurological mechanism, brought about by adolescent development, that could explain much about adolescent behaviour and tendencies. For instance, adolescents show greater levels of novelty seeking, they love trying new things, even if those things are of questionable legality. They also seek out more interactions with their peers, friends, and others like them (teenage 'gangs', anyone?). By contrast, teens regularly fight and argue with their parents. All of this is worsened by adolescents having greater tendencies towards risk-taking. How these things are expressed varies considerably; you could go travelling to experience new things, new people, and independence, or end up dabbling in illicit substances or underage drinking with like-minded friends. Personality, circumstance and background are major factors here, and so on.[40]

Some say teenagers are more impulsive, but that's wrong; impulsivity is doing something without thinking about any potential consequences, whereas with risk-taking you do think of the (likely) negative outcomes *but do it anyway*. The distinction is important; children can be impulsive, and do things like try to eat dangerous objects or stick their fingers into power outlets, because they don't know better. But studies suggest that, in hypothetical scenarios at least, adolescents are perfectly capable of rational thought, predictions, and appropriate decision-making.[41] It's just that, in real situations, in 'the heat of the moment', they tend not to, and are far more vulnerable to emotional influence rather than logic and reason. Professor Casey and co. argue that this is due to different rates of maturation between the brain's prefrontal cortex and limbic system regions.

Our brains are still developing during our teenage years, but it's a different sort of developing to that seen during childhood. The different parts of the brain are formed and doing their thing, but now refinement, efficiency and specialisation are paramount. Put simply, during childhood, all the parts of our brain are saying, 'What is my job, exactly?' During adolescence, it's more: 'I know what my job is, but how am I supposed to do it?'[42]

Adolescent maturation brings about changes, increased activity and efficiency in those areas responsible for emotion, pleasure and happiness that we've covered extensively, namely the subcortical limbic systems, like the amygdala, and basal ganglia which includes many regions like the striatum and nucleus accumbens. These regions are also responsible for reward anticipation, and, via the action of dopamine neurons linked to behaviour-controlling regions like the prefrontal cortex, also govern and induce reward-seeking behaviour. In other words, they make us want things, and compel us to get them.

As adults, we're not as beholden to these powerful but primal influences; as we've seen, our rational, impulse-regulating prefrontal cortex can assess the long-term outcomes of emotion-driven, gratification-seeking behaviour and say 'no, that's not a good idea', so override it. The problem is, during adolescence, the emotional, reward-seeking regions mature faster than the prefrontal regions. You'd maybe expect as much, with emotion and reward regions being more established and less 'complex', but it means that, for a prolonged period, they have greater influence over our behaviours, while the more disciplined parts of our brain are still developing, still figuring themselves out. In many ways, it's a lot like the emotion versus thinking issue that children deal with, but more complicated. It's not newly forming parts of the brain just trying to shout over each other, it's subtler, more sophisticated. Less *Jerry Springer*, more *Game of Thrones*.

Imagine someone riding a horse. The horse does a lot of the work, but the rider's in overall control. But then the rider realises she doesn't know where she's going, so stops to consult a map, and lets go of the reins, meaning the horse is in charge. The rider then finds herself knee-deep in a stream, or in the middle of a field. In this example, the prefrontal cortex is the rider, the subcortical limbic systems the horse. Essentially, letting the less sophisticated elements assume control means you end up in unhelpful places and situations. As teenagers often do.

This explains a lot about adolescents. Sure, they are perfectly capable of thinking clearly and calmly in hypothetical scenarios, where their emotional responses aren't being engaged. But most real-life situations have a strong emotional component, which would strongly influence adolescent behaviours and decisions, given the arrangement of their brains. You ask a teenager 'Do you hate your parents?' and they'd probably say no, of course not. But, if their parents say they can't go out or have the latest smartphone, they may well scream 'I hate you!' because *in that brief moment*, they sort-of do; a fleeting yet intense emotional burst can be more powerful than logic and reason to an adolescent brain. Then they'll slam some doors, as is traditional. This also explains adolescents' blasé attitude to risk; their brains are more susceptible to emotional drives, immediate stimulation, and gratification, while being simultaneously less susceptible to long-term consequences and rational thought. *Of course* you'd see more 'risky' behaviour.

The maturation of the limbic system and reward pathways

also means things that previously made us happy abruptly lose their potency, so what we once thought of fondly now seems childish and embarrassing. Increased efficiency and influence of the striatum and amygdala, with all their social functions, could lead to greater need for companionship and acceptance, and a greater desire for high social status, hence the classic teenage obsession with being popular and 'cool'. Of course, a sudden desire to explore, indulge and be 'top dog' regardless of risk is not something your parents are likely to be thrilled about, so they'll inevitably thwart all your new desires. Even if this is well meant, one's basic needs and drives being denied causes anger and stress.[43] And as teenagers are more sensitive to stress and anger, so they will more often lash out at parents and authority figures. Overall, those who once provided stability and security are now perceived as barriers to growth and self-discovery, meaning they're resented rather than appreciated.

While these unhappiness-inducing behaviours may look like unfortunate quirks of development in the ever-complex human brain, they seemingly exist for a reason. Rats and primates, both social creatures themselves, also show similar or analogous behaviours during their adolescent stages,[44] suggesting they are indeed advantageous, and here's one explanation as to why.

When we become sexually mature, ideally (from an evolutionary perspective) we'd go and seek out potential mates and attempt to 'woo' them. A heightened sex drive coupled with desires to meet new people and take risks would be very helpful here. Working against this, though, would be existing preferences to stick to the safe and familiar and avoid responsibility, by remaining close to your family group. However, regularly arguing with or resenting your parents means you're more likely to strike out alone, improving your chances of mating and later success.

Not every teenager does this, obviously. We all mature in our own way and at different rates. Some teenagers clearly maintain focus and responsibility throughout their adolescence, but it's likely to be harder for them to do so, on a neurological level, than if they were adults. And perhaps the main problem with happiness for adolescents is not the neurological changes occurring, but the fact that modern society largely fails to account for these in any meaningful way. Adolescents and their newly developed brains are sexually aware/motivated; they want independence and control over their lives, to experience new things and meet new people. However, society has many restrictions in place, be they age-related, financial or cultural, to prevent them from doing much of this. It's perhaps understandable, if not necessarily acceptable, that adolescent frustrations can end up being taken out on the society that causes them, via vandalism or other illegal activities.

It's a cruel irony that adolescents are expected to behave like responsible adults while having the rights of children (e.g. in the UK they must choose study subjects that will define their whole lives at around fourteen, but can't be trusted with a beer until eighteen), when, neurologically at least, they're neither of these things. They're adolescents. Perhaps until the wider world recognises and meaningfully accommodates this, adolescents are unlikely to be reliably happy, as their needs and desires are, suddenly, incompatible with much of the world they live in. Until then, their happiness may depend on whatever indulgences are afforded to them to vent their pent-up aggression, stress, and need for novel stimulation. Eye-wateringly graphic and violent video games could be one such outlet, particularly the modern online types that allow you to connect and converse with your peers, and best them in contests.

It could be the case that, far from encouraging bad behaviour and corrupting fragile minds, violent video games are the only things keeping some teens happy and relatively balanced, and without them they really would be trouble. This isn't something that many scaremongers want to hear, but hey, don't shoot the messenger. Maybe you should learn some self-control?

## A grown-up approach to happiness

So, after adolescence, adulthood. Freedom and independence, yay! Self-reliance and responsibility, boo! It is a mixed bag, in fairness.

Like the onset of adolescence, the question 'at what point do you become an adult?' is also tricky to answer. 'Biological adult', in the scientific literature, usually means an individual who has reached the stage of sexual maturity. But, for humans, this would mean the *onset* of puberty, so are eleven-year-old children technically adults? Most would dispute that. It may work for other species like rodents who only live a few years, but long-lived humans with their weirdly prolonged youth phase[45] decided this wasn't right, so developed the concept of 'social adult', where the conventions and laws of society decree you to be an adult because you have reached a required age or milestone. These vary considerably from society to society.

In neuroscientific terms, the point where the brain has officially 'finished' developing and maturing is also tricky to isolate. We've seen that different parts of the brain mature at different rates already, and there's a lot of evidence revealing that some carry on doing so through our twenties, with areas like the corpus callosum (the 'bridge' between the two hemispheres ) and frontal lobe areas for important executive functioning and conscious control showing signs of continual development roughly up to the age of twenty-five.

So, let's say, neurologically, we're 'fully adult' by age twenty-five. Assuming a typical lifespan of seventy (although this seems to be increasing all the time[46]), that still means you're an adult for most of your life by some considerable margin, so it's your adult brain that will decide if you experience 'lasting' happiness. Your personality, temperament, likes and dislikes, abilities and inclinations, are all essentially baked into your brain during the development stage, and are the things that determine what makes us happy, to what extent, and why.

But are they set in stone? You'd perhaps expect the adult brain to be a lot more 'fixed' than earlier iterations, and in many ways, it is. When I showed my young children a smartphone or tablet computer, in barely five minutes they were using them as well as me, even though I, a child of the eighties, still consider touchscreens borderline sorcery. Comparatively, if you've ever tried to teach an elderly relative how to use such things, you'll know it can be something of an uphill struggle.

For many years, it was indeed widely believed that the adult brain was essentially 'set', with all the neurons and major connections we'd need. Sure, we learn new things and update our understanding of things all the time, meaning new connections are regularly being formed and turned over in networks governing learning and memory.[47] But in terms of overall physical structure and major connections, the stuff that makes us 'what we are', the adult brain was long thought to be 'done'. However, in recent years there's been a steady stream of evidence revealing that the adult brain *can* change and adapt, even create new neurons, and experiences can still reshape the brain, even as we head into our twilight years.[48] Consider the taxi driver study from chapter two, where constant driving and navigation of chaotic London streets leads to increased hippocampus size, revealing the adult brain structure is somewhat malleable. One thing seems clear, though; it takes a lot more effort and time to alter an adult brain, compared to a younger one.[49]

Intelligence, for example, a product of the efficiency and intricacy of numerous brain connections, is a lot harder to alter as an adult.[50] It *can* be done, but only with considerable time and effort, for very little notable gain. There are plenty of products and games out there that claim to 'boost your brain power', but they're misleading at best. Doing crosswords and number games every day will certainly improve your abilities, but only with regards to doing crosswords and number games, because the brain is a lot more complex and versatile than that when it comes to intelligence; you're just enhancing one particular facet of a sophisticated system. It's like a general discovering his army is only half as big as he'd like, so he sends one soldier to the gym for a month to get bigger and stronger. At the end of that, he's got a more powerful soldier, but his army's no bigger, so it's not really solved the initial issue. It doesn't mean what you've got is bad or ineffective, and it can be used in very impressive ways still, it's just hard to change the basic elements.

So yes, the adult brain *can* change, it just takes considerable time and effort when compared to younger brains. It went through all the tumultuous development for a reason; you can't blame it for not wanting to do all that again.

What makes an adult brain happy? That can't be answered in any succinct way, sorry. Everything covered in the previous chapters applies to adult brains, but how much or little they apply to yours, that's for you to judge. No two people are alike, and what makes them happy, be it a nice home, family and friends, love and sex, laughter and humour, sporting achievement, a successful career, vast wealth or fame, creating masterpieces or just reading a book, depends on who they are and how their brains respond to such things. Most people would be made happy by many/all of those, but at different times and for different reasons. Because of the way we've evolved and the world we've created around us, there are just so many things that can make a modern adult brain happy.

That's lucky really, because if there's one factor that affects the brains of all people, it's stress. Stress chemicals like cortisol, the threat detection circuits of our brain, the amygdala's fear-producing processes, the fight-or-flight reflex, are all ancient and deeply entrenched elements of our brains that mean we react strongly to anything potentially dangerous or threatening. However, one downside of the vast expansion in human intellect is that it's now a great deal easier for us to experience stress, because we're 'aware' of far more dangers and threats. For a simpler animal, stress could be caused by things like 'I'm sure there's a predator around here somewhere' or 'It's been a while since I last found food'. Humans have a much richer selection of stressors: what if I lose my job? Do my in-laws like me? Do I have enough spending money?

Am I too old to start a family? What if I never visit Paris? How do I help the victims of that tragedy overseas? Why does my chest hurt? The economy's not looking too good. My Wi-Fi's gone down! And so on.

Being an adult means stress. Before, your parents made all the important decisions and paid for things; now it's on you. Sure, you can go out when you like, eat what you like, meet who you like, but you also must fund these things, and look after your own long-term health, and decide whether or not these people you're meeting are trustworthy or safe, because that's often not guaranteed. Children are, usually, largely shielded from consequences, and adolescents seem less bothered about them in pursuit of more immediate gratification, but adults often have no way around them. With so many decisions and actions that could feasibly come back to bite you, adulthood is a stressful time. And that's without being responsible for anyone else's wellbeing, which many adults are.

This isn't great, in terms of health. Constant, chronic stress is a huge problem in the developed world, because it has so many health consequences,[51] and we've built an environment where regular stress is just a part of life. There's only so much stress a brain can endure before it is pushed beyond its limits, but this varies from person to person. Because of this, in 1977 psychologists Zubin and Spring came up with the stress-vulnerability model of mental illness.[52] It's a straightforward way of modelling the fact that the more vulnerable someone is to stress, the less stress is needed to cause a breakdown and develop some mental-health problems. Those with harder lives, more difficult situations, or prior histories of poor mental health, will have fewer brain resources available to deal with any further stress that occurs, whereas someone skipping through life whistling all the while could probably shrug off a brief period of hardship, should such a thing happen.

And that's where the importance of happiness comes in. Studies reveal that things that make you happy, that increase activity in the reward pathway, seem to directly combat the physical effects of stress in the brain and body.[53] Besides being enjoyable anyway, the pursuit of happiness may well be what's needed to keep your adult brain's level of resistance to stress topped up as much as possible, to better help you deal with the issues and hazards that life inevitably throws your way.

It's not *that* simple, obviously. Nothing involving the brain ever is. Things that make us happy can end up causing stress, and vice versa. Indulging in delicious, high-calorie food is immensely enjoyable and known to reduce stress, but too much of it means you gain weight and your health suffers, which causes stress. Travelling to exotic locations is something that reliably makes people happy,[54] but costs a lot of time and money, which you may need to avoid stress later. Conversely, putting yourself through the stress of exams, training, dieting, etc. can mean you achieve longer-term goals which make you happier later. It's a complex, confusing system, and one we're working out as we go along, to the extent that life and circumstances allow (because they often don't).

The overall point is, for the adult brain, experiencing happiness may well be more of a necessity, rather than an indulgence. Of course, while it's easy for me to say that it's important to make sure you're happy for the wellbeing of your brain, people don't exist in a vacuum. We're all part of one big community now, or multiple small ones, and we've seen how the human brain craves the approval of others. Unfortunately, what makes you happy may not be approved of by others, and what others assume, *insist*, will bring about happiness, could well leave you cold. We saw this in chapter five with the 'relationship escalator', where social norms and expectations mean people in the Western world have a fixed and rather narrow model for how romantic relationships are supposed to work – a model that an increasing number of people are realising doesn't accommodate what makes them happy. Societal expectations are powerful things, and they can easily get in the way of individual happiness.

For instance, one of the major sources of both stress and happiness for the adult human is having children. Bringing a life into the world has massive consequences for your own, and wily evolution has instilled in our brains numerous traits to encourage it, like a tendency to feel affection and happiness and to be motivated towards caregiving behaviour by anything that even resembles a human baby,[55] hence we fill our homes with puppies and kittens and other pets with big heads and eyes and child-like personalities. And if they're our own offspring, well, empathy and bonding and protective instincts are off the map. Obviously, when we're of an age, we want to have children.

Except, some people don't. Be it due to quirk of brain chemistry, health concerns, environmental influences, or just thinking about it and deciding it's not for them, many people don't have children, and never intend to. They know what does or will make them happy, and it isn't reproducing.

One such person is UK technology journalist Holly Brockwell, founder and editor of the female-oriented tech and lifestyle site Gadgette, among other things.[56] But outside of that, she's also caused a stir via her candidness about her one-woman campaign, since successful, to convince the NHS that she should be surgically sterilised.[57] This caused her to experience a significant backlash and she gets criticism and condemnation from random strangers online to this day. Why, though? Why would anyone else care what a woman they don't know does with her own body? They'll never meet the theoretical children she'll not have; it's actually cheaper for the NHS overall compared to supplying a lifetime of contraceptives or handling any birth(s) she would have; vasectomies and abortions are permitted; and with seven billion humans and counting currently occupying the planet, I don't think the species is in any danger of dying out soon. So what's the problem? These arguments eventually won the NHS guardians over, so why do people still get in her face about it? I figured I'd ask Holly directly: why was she so certain she didn't want children?

'I never felt the need for children, ever. But I was told when I was younger that this was something I'd "grow out of", and I believed that people knew better than I did about it, so I believed I would eventually want kids. So for a while my vision of a future for myself included children, though I always felt scared about it. Eventually I realised that people didn't actually know better after all, and that my feelings of not wanting babies were completely valid and actually quite common. It turned out, in fact, that my own mother hadn't wanted kids, but with times being different, she hadn't been able to make the same choice I have.'

Times are indeed changing with regards to the greater degree of choice and autonomy young people are growing up with, which is probably good for overall happiness, although a lot of people do seem to be scared by that. Holly told me that she'd even met men supposedly only interested in casual dating who were outraged by her affirmed desire to not have children, with one guy walking away less than three minutes into a speed date because of it. A speed date! And while I'm not the sort to reduce an adult woman to their physical attributes, if I had to provide a description of Holly's appearance the word 'unattractive' would not appear in it anywhere. That speed date guy sounds like quite the mother-lover.

Holly even does her best to make it clear that while she doesn't want children, she doesn't *hate* children. She loves her nieces dearly. She just doesn't want any of her own.

'It's not aversion to children themselves, but what I know my life would be like if I had them. I know myself well enough to know I'd be very unhappy if I went down that path. Of course, if I somehow ended up with a child I'd like to think I would love and care for them the way my mum did with me, but in the same way you can do decently well at a day job that leaves you watching the clock, you'd still be happier in your dream career. For me, the ideal life doesn't include kids of my own.'

It seems the mere notion that a woman wouldn't want children is upsetting for many. Maybe it challenges a core belief, that women love children, that some people's worldview appears to be based around?[58] It's nothing to do with them, yet we've seen how people can turn on each other in pursuit of their own happiness. Maybe some are thinking they're even helping her, like the Christian street preachers haranguing the passing heathens? Who knows. Point is, when you're an adult it's important – maybe even necessary – to be happy, but sometimes your happiness is contingent on the acceptance of others. That's the problem with adulthood, I guess; for all that our brains are 'mature' now, we're all just figuring things out as we go along, in a world that we're constantly changing just by being in it. Sounds like a stressful existence.

## At the end of the day

I said earlier that it takes a lot of time and effort to change an adult brain. Effort is something we invest carefully given how our brains process it, but time? Time keeps coming, whether we like it or not. Amazing, baffling, incredibly complex and alarmingly powerful as the human brain may be, it is still just a biological organ, part of the body. And the body ages. Wear and tear gradually takes its toll, and, as you might expect, this can make a big difference to our happiness.

Even if our brains were somehow invulnerable to the physical effects of age but our bodies weren't, this would still be certain to make us less happy eventually. Our bones and muscles weaken, our joints and digits stiffen, our eyesight and hearing fade, our heart weakens, our arteries harden, our libido wanes, and so on and so on. All of this can make you unhappy, simply by compromising your ability to do/experience the things that usually give you pleasure. You enjoy hiking and visiting art galleries? Tricky to do those if your hips are giving you grief and you need to get your cataracts sorted.

Heck, it doesn't even need to be anything so pronounced. Maybe impressing others with your good looks is what makes you happy? Losing your hair or having it go grey, along with reduced skin elasticity causing wrinkles, is going to hinder that, particularly in our youth- and image-obsessed world. At least, I assume it would. As someone whose hair started receding at age eighteen, this isn't a dilemma that will ever concern me.

And that's not to mention the possibility of becoming seriously ill, the odds of which go up and up as we age. Sure, most major illnesses are unlikely to happen to you, but the longer you live the more susceptible your body becomes, and the more chances rogue genes or unseen environmental hazards have to work their evil way on your bodily systems. You keep rolling the dice day in, day out, you're bound to hit snake eyes eventually. Many serious, debilitating conditions often end up with the patient experiencing depression or similar mood/anxiety disorders as well,[59] because why wouldn't this be the case? Of all the things that could cause stress and max out the brain's coping mechanisms, 'terminal illness' must be top of the list, or close to it.

Age also has other, less physical consequences that can lead to unhappiness. One is that for most of your life you may have had a goal or ambition you were working towards, but at a certain point that's no longer the case. You're either too old to do it now, or you have achieved your ambitions, so there's no need to do them again. As Kevin Green astutely pointed out, it may seem idyllic to be free of the day job, financial obligations and any responsibility, and for many people the resultant freedom no doubt is brilliant, allowing them to do all the things they've always wanted to do. But, the rather abrupt loss of routine, of responsibility, of *purpose* can be genuinely debilitating, leading to all manner of psychological consequences like depression,[60] which impact on our physical wellbeing. This is no small thing when you're older.

It's also worth remembering that nobody lives forever, and the older you get, the more likely it is that you'll experience those close to you, be they friends, family or partners, passing away. Grief is a purely natural but very potent emotion, and can take a long time to adjust to and get over. Indeed, some can find it impossible to move on, essentially becoming isolated and 'addicted' to the memories of the departed, to the point where therapeutic intervention may be necessary.[61]

And to top it all off, while we're often told to 'respect our elders', our society doesn't always practise what it preaches. Older people are often marginalised and ignored in the mainstream media, and even by their own families, who now have their own lives and goals and responsibilities to deal with. Taking care of an increasingly frail parent or relative is a big responsibility, one that gets increasingly demanding over time. Coupled with close family often being spread over wider areas due to readily available transport and the nature of modern work, the end result can be older people ending up neglected and largely forgotten, and their increasing fragility means they can't do much about this. They end up lonely, a major issue with our ever-expanding elderly population,[62] and require help and assistance with their daily lives, leading to loss of autonomy, more stress, further unhappiness, and the vicious circle continues.

And all that's assuming the ageing brain remains the same. It doesn't. The brain is the body's most energetic organ, and all the exotic and constant processes it engages in take their toll on its very structure. Age hits the brain in many different ways,[63] but particularly relevant ones are a depletion of the dopamine and serotonin systems. Dopamine is crucial for experiencing many emotions and the functionality of the reward system, so this would obviously impact the ability to be happy. Serotonin is a key transmitter for mood stability, and also influences the sleep cycle.[64] Older people tend not to need as much sleep, but this can lead to cognitive and mood issues of its own. Even assuming there's no neurodegenerative disorder occurring leading to things like dementia (always a risk with advancing age[65]), the aged brain is still less flexible, less efficient, less quick on the uptake as it was in its prime, even less able to process emotions as effectively as it once did,[66] which obviously affects your happiness. That's just how it goes; entropy gets into everything eventually.

I realise this is a pretty grim picture, so apologies for that. However, I figured it would be best to get the bad news out of the way first, so it's more rewarding when I explain what can be done to prevent or combat all this doom and gloom. Everything I described above is only 'inevitable' if no effort is made to address the effects of ageing, and thankfully there are many options for doing so, some of which have been gifted us by evolution itself.

Firstly, these days there seems to be a constant stream of studies showing that regular exercise is a reliable factor in staving off the negative aspects of old age.[67] This makes perfect sense, because as stated the brain is a bodily organ, and exercise increases the metabolism and improves the health of your heart and associated systems, meaning more blood and nutrients gets pumped throughout the body, meaning in turn that the brain has greater reserves of minerals and energy to keep itself active, which is all for the good.

Indeed, an active brain is a healthy brain, and those with higher educational levels seem far more resistant to cognitive decline, even if the physical mechanisms that would lead to this are advanced.[68] Thankfully, you're never too old to be educated; barring some major disruption to the memory system (which is, admittedly, a possibility in old age, via things like dementia), your ability to learn new things persists throughout your life. Taking classes and the like may not have much career benefit when you're retired, but that doesn't mean there's *no* benefit.

Many cities around the world have now introduced playgrounds designed for the elderly,[69] so they can get more exercise, thus improving their health and wellbeing, but hopefully in a fun and interactive way. Nothing wrong with feeling like a child again, if it makes you happy.

That's another thing: nostalgia. Older people are often thought of as looking back fondly at the past, assuring everyone that things were better 'in their day'. In a way, this is perfectly logical; you would prefer to think about a time when you were in your prime, rather than the present where you're aged and less capable. This can sometimes go too far though, when people's recollections, distorted and coloured by the brain's optimism bias when it comes to memory,[70] interfere with their modern-day existence. For years, many psychologists and therapeutic types genuinely considered nostalgia to be something of a disorder,[71] or at the very least a negative cognitive behaviour, distracting from the now to focus on an inaccessible, exaggerated point in the past.

Now, however, evidence suggests that nostalgia (at any age) is actually a very *positive* process, and can make us more motivated, more social, more optimistic, all things that boost our wellbeing and happiness.[72] The logic is that thinking regularly about your past *which was good* means you retain an awareness of your own achievements and abilities, and can more easily accept that positive things can and do happen, which just makes you feel better. It seems that nostalgia isn't so much mourning for what's lost, but more appreciating what you've achieved. It's mentally polishing your trophies, not pining over failed relationships.

Obviously, this can go too far; older people voting en masse to recreate the quasi-fictional romanticised world of the past doesn't really do much good for anyone (see 'Brexit'). It makes you long for the days when nostalgia was considered a problem, which is about as self-defeating as it gets.

Finally, the main way in which we can stave off the downsides of ageing on the brain seems to be sociability, that common element in so much of our happiness. It's isolation and loneliness that seem to be the most harmful (non-physical) factors in the psychological wellbeing of the older generation, so anything that can safeguard against these scenarios is bound to improve happiness.[73] Hence the frequent refrain of an older person who 'just wants someone to talk to': we're humans, we've evolved to need to be around others, and nobody has ever got so old they've evolved into something else.

In a way, that's sort of why people get so old in the first place. Humans live way longer than similar species, and we persist long after our physical or reproductive usefulness has peaked, which isn't very 'selective' if you think about it. There are many theories as to why we ended up so enduring, but one factor seems to play a big part, which is the positive influence of *grandparents*, on the survival of young ones and the community at large.[74] Older members of a primitive human community may not have been much cop at hunting or the physical stuff, but they were still perfectly capable of looking after the babies and children, and didn't need to spend time pursuing mates or any of that other exhausting stuff. The children were cared for, learned wisdom was passed on directly, extra hands were available for the day-to-day stuff . . . The advantages of keeping your old folks around are many and varied, it seems.

Being a grandparent gives a new set of (ideally less demanding) responsibilities and focus to the older generation whose own children are grown and independent now. It's no wonder many of them are unashamedly eager for grandchildren, like my folks and in-laws were. But it's a two-way relationship; grandchildren get cared for, grandparents get to care for them. Everybody wins.

Of course, not everyone is fortunate enough to have close enough family (emotionally or geographically) for this to be an option. But it does seem that keeping in touch with others is what is required to stay happy and stave off the inevitable, for the ageing brain.

And 'inevitable' is a loaded word there. No matter how much effort we put into keeping sharp and cheerful, we'll stop eventually. Because we'll die. There's no getting around that, sorry. Every human and their brain are finite. We will ultimately expire, we just don't know how or when. And that's helpful. It's maybe the one instance where uncertainty actually *decreases* stress and keeps us upbeat.

But, that's the downside of the human brain; it being as powerful as it is, allowing us the understanding it does, and having created the advanced medical science we take for granted nowadays, it is now entirely possible to know, roughly, when you're going to die. We can now diagnose terminal illnesses and provide a prognosis, so the afflicted are palpably aware of how much time they have left, at least to within a reliable ball-park figure.

What must this do to a brain, and one's happiness? How do you come back from that, mentally? It's always baffled, alarmed and amazed me in equal measure, from a psychological perspective. And I'm not talking about the mystical, theological elements here; that's for philosophers and relevant scholars to deal with.

One person who has no time for such considerations is Crispian Jago, avowed atheist and skeptic, vinyl record lover, Cornishman and wit. I met Crispian in his capacity as organiser of the Winchester branch of grassroots rationalist organisation Skeptics in the Pub (I founded and ran the Cardiff branch for several years). However, in 2016, he discovered he had terminal, incurable cancer, and was given eighteen months to live.[75]

Now aged fifty, a year into his eighteen-month expectancy with his signature ruddy red hair and beard now turned snowy white by the chemotherapy, I asked Crispian about how he's dealt with having to reconcile happiness with impending mortality.

'When I was told that my cancer had come back and spread and that it was inoperable and terminal, happiness is not the emotion that best described how I felt. I felt cheated, especially after working hard and getting myself in a comfortable position ready to enjoy my retirement. I never felt particularly angry, just a sense of misfortune and, admittedly, self-pity.'

Yes, Crispian is exactly the sort of classic British cliché who feels the need to apologise for experiencing self-pity when dealing with a terminal illness.

After several months of despondency and understandable upset, in the last six months he's felt his happiness return. He's had to take most of the year off work, so has ended up going on the sorts of trips he was planning for his retirement. He's seen his teenage children get into good universities and knows they'll be OK and have good lives. And, as he points out, 'I've been able to watch all five days of the first test at Lord's without work getting in the way.'

One interesting observation Crispian shared was that, with his chemotherapy, he has good days and bad days. Some days he feels rotten after it, others he feels fine. Being the rational, analytical sort by nature, he spotted that the good days seemed to coincide with times when he met up with old friends and well-wishers, who have turned up a lot more since his diagnosis. Spotting this trend, he's made every effort to surround himself with those he cares about, as often as possible, and it seems to have done the trick.

'Having terminal cancer has demonstrated to me, beyond doubt, that I am in fact loved by a great many people, friends and family alike. People often don't bother to say this to you when they don't think you're about to die. However, if you are, they seem to make a much greater effort to tell you what you really mean to them. This also makes me very happy.'

It would seem that enjoyable interactions with others, and the mutual love and approval that comes with this, is what's had the biggest impact on Crispian. If there's a more powerful endorsement for the power of positive relationships with other people to make you happy, I'm afraid it'll take someone far more capable than me to find it. Many people might turn to God and spirituality at such times, but avowed atheist that he is, Crispian has no time for that. He insists that it's helped to have a clear mind, and not be worried by eternal judgement and all that stuff.

'After several sorrowful months following my prognosis, happiness has unexpectedly returned thanks to friends, family, relaxation, happy memories, no regrets and critical thinking.'

I could say more about this, but what exactly could I add? Despite his situation, because of all these positive aspects he's embraced, Crispian sure seems to have a happy brain. Which is what I've been looking for all this time.

# Afterword

You know when you set off on a long car journey, after spending hours packing and checking everything, and before you've even reached the end of your road that nagging voice in your head starts off, insisting that there's at least one important thing that you've forgotten, or not done, or shouldn't have done? Did you leave the central heating on? Is there enough food for the goldfish? Are you *sure* there's a key under the doormat for the house sitter? It looked like the bedroom was on fire when you checked, should you have done something about that? And so on.

That's the exact feeling you get when you finish a book like this, except multiplied by a million. Although in my defence, I *know* I've definitely forgotten to include numerous relevant things. I was just recently telling a friend I'd pretty much wrapped up my research into happiness, and she asked what I thought about those international surveys that determine which countries are the happiest. This resulted in me pausing for an uncomfortable length of time before uttering a guttural scream while repeatedly dragging my fingernails down my face.

I'd also like to take this opportunity to publicly apologise to that friend, whom I've not seen since.

I've since looked up those international happiness surveys, the most robust of which seems to be the OECD (Organisation for Economic Co-operation and Development) 'Better Life Index', which was launched in 2011 after a decade of work and research. It takes the form of an interactive tool that allows countries to measure how good the lives are of the average citizen, and work out how 'happy' the country is overall. I was bleakly considering rewriting whole chapters to accommodate this, but then I took a look at the categories the index measures and via which it assesses the average person's wellbeing. These are housing, income, jobs, community, education, environment, governance, health, life satisfaction, safety and work–life balance.

Looking back, I *have* covered all of these things, one way or the other. Some I've tackled directly and at length, others can be explained in the context of the neurological properties I've explained elsewhere. And in fairness it was never my plan to actually measure happiness, I just wanted to see what makes our brain happy, and why. The fact that a multinational years-in-the-making project can arrive at similar conclusions as me and my repeated blundering into the scientific literature interspersed with asking various people 'what do you think?' suggests that there may be some merit to what can be found in the previous chapters after all.

There are undoubtedly other things I've not included, though. Why do things like sports – often aggressive physical competitive acts – make us happy, whether we're playing or watching? Why does a family gathering, supposedly a happy event, so often end up filled with stress and bitter recriminations? Does all the sex stuff still apply if you're homo- or bisexual? And what about transgender people? Or those with mental-health issues of some description? Where does that all fit in with how our brains make us happy? One of the reasons I couldn't answer all these questions is because I simply didn't have room; this is a huge subject matter after all, and it encompasses way more than one reasonably sized book could ever hope to include. Other times, I was simply restricted by the available scientific data. Society and what's considered 'normal' can change quickly, while scientific practices, painstakingly worked out over the decades, do not. It's hard to answer these sorts of questions objectively if the available evidence doesn't consider them.

But, what *did* I find? After all this, what's the secret to lasting happiness, based on what I now know about how the human brain deals with it? Well, it should come as no surprise to anyone who's read everything before this point, but it doesn't look like there is one. Happiness isn't stored in the brain like gold bullion in a treasure chest, just waiting for someone with the right key to turn up and spend it. The human brain never has been, and never will be, as simple, straightforward and consistent as that. It turns out there are plenty of things that do stimulate our brain in just the right way to make us happy, but each of these comes with caveats and limitations.

For instance, it's hard to be happy without a home to call your own, a place that provides a reliable safe place where you can shut out the big scary world and regain control over your surroundings. But, it has to be the right *kind* of home, it has to tick enough of our boxes so that we feel comfortable there, we can feel it truly represents us, meets our (highly individual and often arbitrary) requirements, and so on. And, as important as it may be in so many ways for your happiness and more, your home is often determined by more external factors, like work and family concerns.

Our work is something else that can make us happy, as long as we maintain a decent work–life balance. However, what counts as a valid 'balance' varies considerably from person to person, and the nature of the working world means that while our jobs can be immensely satisfying and rewarding in ways our brains readily respond to, they can also be dispiriting and dreadful, triggering the stress and negative emotional reactions that our brains are only too willing to deploy at a moment's notice. Some people, for many reasons, end up with brains that like to work as much as possible, whereas others become miserable if they have to do more than the bare minimum. On top of all this, the nature of much modern work has many different effects on us and how we perceive our place in the world.

Much of this comes down to money, of course. We need money to live, and we need to work to get money. It seems our brains are made happier by financial reward as a result, but only up to a point. If you end up having more money than you *need* to ensure your survival in our complicated world, then the relationship between money and your happiness starts to blur and shift, and other factors can take priority. Your brain is perfectly capable of recognising that your financial situation has changed, and with this can come a whole suite of new issues and priorities that determine your success. Or failure.

Because we all want to succeed, at least in some shape or form. Because we want, nay *need*, the approval of others. We're a social species; many theories argue that our ability to make friends and interact with others is what drove our brains to become so powerful in the first place, and an incredible amount of our brain's abilities and functions are geared towards enhancing our communication and interaction with those around us. As a result, approval from others, in whatever form it takes, is highly valued by our brain's underlying systems, and makes us happy as a result. It's tempting to say that the more people like us, the happier we are, hence fame is something so many people crave.

Again though, it's not that simple. Like with money, once you get to a certain point with fame, it starts to become less potent and rewarding, and it's the respect and approval of those you're closest to that makes you happier. Without this, you can end up going 'off the rails' somewhat.

Even if you don't want fame, you are very likely to want the approval and affection of a special someone, in both the mental and physical sense. Love and sex are huge interconnected factors in our happiness, for all that they're often treated very differently. So fundamental are these to our day-to-day existence, and so extensively have we evolved to obtain them, that they have many significant (and often destabilising) effects on our brains, altering our behaviours, our thinking, even our very perception. Much of this makes us happy, sometimes euphoric, but it's also messy and complicated, and when people see obtaining love and sex as goals to be met, this can make you unhappier in the long run: they're *part* of life, not the endpoint. There's no finish line to cross or 'game over' sign that flashes up when you settle down with someone. Life goes on, and so do you. As happy as it can and does make us, when we regard finding love as some sort of quest for treasure, we risk distorting how the brain works and what the whole point of it is.

Similarly, laughter and humour are fundamental, pleasurable and widespread elements of happiness. Everyone enjoys them, utilises them, seeks them out, because they affect us in many ways and have developed several diverse functions in the modern human brain. But, as great as humour is and as happy as it can make us, it seems that basing your life around it isn't a guarantee of lasting and reliable happiness. It can, in certain circumstances, do more harm than good.

Not that doing ourselves harm, or causing it to others, is an automatic barrier to happiness. Thanks to the myriad properties of our baffling brains, there are many cases in which what makes us happy also causes us damage, or compels us to inflict it on others. The basic assumptions and mechanisms of the human brain haven't caught up with many features of our advanced, complicated modern world, and this means we often end up finding happiness with things that fly in the face of survival instincts or social harmony. You know, the stuff we're meant to care about.

But, as we've seen, what we care about isn't fixed. Your brain, which is you after all, changes as you age, as you enter different stages of life and development, and these changes can occur at the deepest biological levels of the brain, meaning what made you happy when you were younger will no longer do so a few years down the line.

It would be nice for there to be a convenient take-home message from all of this about how to be happy, wouldn't there? But I'm afraid I can't help you there. All of the people I spoke to, from scientists to superstars, stand-ups to sexpots, millionaires to those facing their own mortality, all have found happiness in their own personal way, via the different paths their lives have taken. If anything, this whole process has made me even more sceptical of those who claim to know the 'key' or 'secret' to lasting happiness. I'm reasonably confident now that there's no such thing, or if there is, it's different for every individual, so pitching the same approach to the whole population is, at the very least, bafflingly naïve. But then, if the advice dispensed by others works for you, have at it. That's the brilliant thing about the human brain, there's hardly anything it won't take on board and react to, regardless of logic or objective reasoning, and this very much applies to how it processes our happiness.

However, if you held a gun to my head and insisted that I identify an overarching theme that connects everything I've found out about how the brain deals with happiness, it's that so much of what makes us happy is dependent on *other people*. Other people share our homes, our jobs, our hobbies; we work to impress them, seek out their approval, their intimacy, their love, their laughter; we gain satisfaction from besting them in various ways, and even when we end up fearing others, we can gain happiness from causing them harm, as unpleasant as that realisation may be for many. Heck, we like other people so much, we even gain tremendous happiness from creating new ones. Unless you don't want to. That's fine too.

I guess it's true what they say, that no person is an island. This is true in the literal sense; no human being is a large landmass surrounded by water, as that's just daft. But even metaphorically, if there was a time in our evolutionary past where humans (or whatever we were then) could exist happily in isolation, those days are long gone. We're a social species, and even if we value our own space and privacy above all else, the knowledge that there are people out there is a comforting reassurance. So much of our existence is based around our interactions with others, and so much of these affect our happiness as a result.

I'm no exception; after all, I just spent many months writing a whole book for the entertainment of complete strangers. And you just took the time to read it. I hope you're happy with it.

# Acknowledgements

If the approval of others is a big part of what makes us happy, then I'm about to make a lot of people very cheerful indeed, because this book would never have happened without them.

A big thanks to my long-suffering but ever-supportive wife Vanita, who kept our daily lives in one piece while I spent endless weeks tearing out what little hair I have left over this whole thing.

Thanks to my beloved children Millen and Kavita. This is what daddy was doing all those weekends I was hidden away in my office.

Thanks to Chris Wellbelove, my agent, who decided to email a mid-level science blogger out of the blue and say 'Have you ever thought about writing a book?' I hadn't at the time, weirdly. And now look what's happened.

To Tash Reith-Banks, Celine Bijleveld, the various James's, and everyone else at the *Guardian* science network for putting my meandering words in front of people to the extent that I get to do stuff like write books now.

To Fred Baty and Laura Hassan, Faber editors extraordinaire and extremely patient types who managed to restrain themselves from explaining to me what a 'deadline' is. Must have been a herculean effort.

To Donna and Steve and Sophie and John and Lizzie and everyone else at the Faber headquarters who somehow managed to turn my endless waffling text into something people actually want to read. Alchemy is child's play in comparison.

And finally, there are so many people who helped with this book, many of whom are mentioned at length in the text itself, but I always feel the ultimate credit goes to the neuroscientists, psychologists and other true scientists, be they the ones I spoke to, the ones whose work I referenced, or just those still out there now, doing the research that's constantly expanding the sum of our civilisation's knowledge.

I've been there, I know what it's like. Trying to truly understand the brain with the resources currently available is like Sisyphus constantly trying to push that boulder uphill. Except the hill's coated in custard. And the boulder's actually made out of bees. Live, irritated bees.

It's not something I ever sought, but more and more over the past years I've somehow ended up becoming a spokesperson for the neuro community whenever the media need something explained. I guess I have a knack for that sort of thing. But I'm fully aware that I'm not the one doing the actual legwork, I'm basically just a messenger.

I just wanted it acknowledged in print that, like any good scientist, I stand on the shoulders of giants, but I'm not actually going anywhere. I just enjoy the view.

# Notes

## Chapter 1: Happiness in the Brain

1. Burnett, D., 'Role of the hippocampus in configural learning', Cardiff University, 2010
2. Arias-Carrion, O. and E. Poppel, 'Dopamine, learning, and reward-seeking behavior', *Acta Neurobiologiae Experimentalis*, 2007, 67(4), pp. 481–8
3. Zald, D. H., et al., 'Midbrain dopamine receptor availability is inversely associated with novelty-seeking traits in humans', *Journal of Neuroscience*, 2008, 28(53), pp. 14372–8
4. Bardo, M. T., R. L. Donohew and N. G. Harrington, 'Psychobiology of novelty seeking and drug seeking behavior', *Behavioural Brain Research*, 1996, 77(1), pp. 23–43
5. Berns, G. S., et al., 'Predictability modulates human brain response to reward', *Journal of Neuroscience*, 2001, 21(8), pp. 2793–8
6. Hawkes, C., 'Endorphins: the basis of pleasure?', *Journal of Neurology, Neurosurgery and Psychiatry*, 1992, 55(4), pp. 247–250
7. Pert, C. B. and S. H. Snyder, 'Opiate receptor: demonstration in nervous tissue', *Science*, 1973, 179(4077), pp. 1011–14
8. Lyon, A. R., et al., 'Stress (Takotsubo) cardiomyopathy – a novel pathophysiological hypothesis to explain catecholamine-induced acute myocardial stunning', *Nature Reviews Cardiology*, 2008, 5(1), p. 22
9. Okur, H., et al., 'Relationship between release of beta-endorphin, cortisol, and trauma severity in children with blunt torso and extremity trauma', *Journal of Trauma*, 2007, 62(2), pp. 320–4; discussion 324
10. Esch, T. and G. B. Stefano, 'The neurobiology of stress management', *Neuroendocrinology Letters*, 2010, 31(1), pp. 19–39
11. Weizman, R., et al., 'Immunoreactive [beta]-endorphin, cortisol, and growth hormone plasma levels in obsessive-compulsive disorder', *Clinical Neuropharmacology*, 1990, 13(4), pp. 297–302
12. Galbally, M., et al., 'The role of oxytocin in mother–infant relations: a systematic review of human studies', *Harvard Review of Psychiatry*, 2011, 19(1), pp. 1–14

13. Renfrew, M. J., S. Lang and M. Woolridge, 'Oxytocin for promoting successful lactation', *Cochrane Database of Systematic Reviews*, 2000(2), p. Cd000156

14. Scheele, D., et al., 'Oxytocin modulates social distance between males and females', *Journal of Neuroscience*, 2012, 32(46), pp. 16074–9

15. De Dreu, C. K., et al., 'Oxytocin promotes human ethnocentrism', *Proceedings of the National Academy of Sciences*, 2011, 108(4), pp. 1262–6

16. Dayan, P. and Q. J. Huys, 'Serotonin, inhibition, and negative mood', *PLOS Computational Biology*, 2008, 4(2), p. e4

17. Harmer, C. J., G. M. Goodwin and P. J. Cowen, 'Why do antidepressants take so long to work? A cognitive neuropsychological model of antidepressant drug action', *British Journal of Psychiatry*, 2009, 195(2), pp. 102–108

18. Jorgenson, L. A., et al., 'The BRAIN Initiative: developing technology to catalyse neuroscience discovery', *Philosophical Transactions of the Royal Society B*, 2015, 370(1668)

19. Zivkovic, M., 'Brain culture: neuroscience and popular media', *Interdisciplinary Science Reviews*, 2015, 40(4)

20. Pearl, S., '*Species, Serpents, Spirits, and Skulls: Science at the Margins in the Victorian Age* by Sherrie Lynne Lyons', *Victorian Studies*, 2010, 53(1), pp. 141–3

21. Greenblatt, S. H., 'Phrenology in the science and culture of the 19th century', *Neurosurgery*, 1995, 37(4), pp. 790–804; discussion 804–5

22. Sample, I., 'Updated map of the human brain hailed as a scientific tour de force', *Guardian*, 20 July 2016

23. Aggleton, J. P., et al., *The Amygdala: A Functional Analysis*, Oxford University Press, 2000

24. Oonishi, S., et al., 'Influence of subjective happiness on the prefrontal brain activity: an fNIRS study', in Swartz, H., et al., 'Oxygen transport to tissue XXXVI', *Advances in Experimental Medicine and Biology*, 2014, pp. 287–93

25. Kringelbach, M. L. and K. C. Berridge, 'The neuroscience of happiness and pleasure', *Social Research*, 2010, 77(2), pp. 659–78

26. Berridge, K. C. and M. L. Kringelbach, 'Towards a neuroscience of well-being: implications of insights from pleasure research', in H. Brockmann and J. Delhey (eds), *Human Happiness and the Pursuit of Maximization*, Springer Netherlands, 2013, pp. 81–100

27. Witek, M. A., et al., 'Syncopation, body-movement and pleasure in groove music', *PLOS One*, 2014, 9(4), p. e94446

28. Zhou, L. and J. A. Foster, 'Psychobiotics and the gut–brain axis: in

the pursuit of happiness', *Neuropsychiatric Disease and Treatment*, 2015, 11, pp. 715–23

29. Foster, J. A. and K.-A. M. Neufeld, 'Gut–brain axis: how the microbiome influences anxiety and depression', *Trends in Neurosciences*, 2013, 36(5), pp. 305–12

30. Aschwanden, C., 'How Your Gut Affects Your Mood', *FiveThirtyEight*, 19 May 2016, fivethirtyeight.com

31. Chambers, C. 'Physics envy: Do "hard" sciences hold the solution to the replication crisis in psychology?', *Guardian*, 10 June 2014

32. Chambers, C., *The Seven Deadly Sins of Psychology: A Manifesto for Reforming the Culture of Scientific Practice*, Princeton University Press, 2017

33. Cohen, J., 'The statistical power of abnormal-social psychological research: a review', *Journal of Abnormal and Social Psychology*, 1962, 65(3), p. 145

34. Engber, D., 'Sad face: another classic psychology finding – that you can smile your way to happiness – just blew up', 2016, slate.com

## Chapter 2: There's No Place Like Home

1. Raderschall, C. A., R. D. Magrath and J. M. Hemmi, 'Habituation under natural conditions: model predators are distinguished by approach direction', *Journal of Experimental Biology*, 2011, 214(24), p. 4209

2. Oswald, I., 'Falling asleep open-eyed during intense rhythmic stimulation', *British Medical Journal*, 1960, 1(5184), pp. 1450–5

3. Schultz, W., 'Multiple reward signals in the brain', *Nature Reviews Neuroscience*, 2000, 1(3), p. 199

4. Almeida, T. F., S. Roizenblatt and S. Tufik, 'Afferent pain pathways: a neuroanatomical review', *Brain Research*, 2004, 1000(1), pp. 40–56

5. Dickinson, A. and N. Mackintosh, 'Classical conditioning in animals', *Annual Review of Psychology*, 1978, 29(1), pp. 587–612

6. Parasuraman, R. and S. Galster, 'Sensing, assessing, and augmenting threat detection: behavioral, neuroimaging, and brain stimulation evidence for the critical role of attention', *Frontiers in Human Neuroscience*, 2013, 7, p. 273

7. Larson, C. L., et al., 'Recognizing threat: a simple geometric shape activates neural circuitry for threat detection', *Journal of Cognitive Neuroscience*, 2008, 21(8), pp. 1523–35

8. Durham, R. C. and A. A. Turvey, 'Cognitive therapy vs behaviour therapy in the treatment of chronic general anxiety', *Behaviour Research and Therapy*, 1987, 25(3), pp. 229–34

9. Szekely, A., S. Rajaram and A. Mohanty, 'Context learning for threat

detection', *Cognition and Emotion*, 2016, pp. 1–18

10. Suitor, J. J. and K. Pillemer, 'The presence of adult children: a source of stress for elderly couples' marriages?', *Journal of Marriage and Family*, 1987, 49(4), pp. 717–25

11. Dinges, D. F., et al., 'Cumulative sleepiness, mood disturbance, and psychomotor vigilance performance decrements during a week of sleep restricted to 4–5 hours per night', *Sleep*, 1997, 20(4), pp. 267–77

12. Agnew, H. W., W. B. Webb and R. L. Williams, 'The first night effect: an EEG study of sleep', *Psychophysiology*, 1966, 2(3), pp. 263–6

13. Sample, I., 'Struggle to sleep in a strange bed? Scientists have uncovered why', *Guardian*, 21 April 2016

14. Rattenborg, N. C., C. J. Amlaner and S. L. Lima, 'Behavioral, neurophysiological and evolutionary perspectives on unihemispheric sleep', *Neuroscience and Biobehavioral Reviews*, 2000, 24(8), pp. 817–42

15. Mascetti, G. G., 'Unihemispheric sleep and asymmetrical sleep: behavioral, neurophysiological, and functional perspectives', *Nature and Science of Sleep*, 2016, 8, pp. 221–38

16. Burt, W. H., 'Territoriality and home range concepts as applied to mammals', *Journal of Mammalogy*, 1943, 24(3), pp. 346–52

17. Eichenbaum, H., 'The role of the hippocampus in navigation is memory', *Journal of Neurophysiology*, 2017, 117(4), pp. 1785–96

18. Hartley, T., et al., 'Space in the brain: how the hippocampal formation supports spatial cognition', *Philosophical Transactions of the Royal Society B*, 2013, 369(1635)

19. Jacobs, J., et al., 'Direct recordings of grid-like neuronal activity in human spatial navigation', *Nature Neuroscience*, 2013, 16(9), pp. 1188–90

20. Rowe, W. B., et al., 'Reactivity to novelty in cognitively-impaired and cognitively-unimpaired aged rats and young rats', *Neuroscience*, 1998, 83(3), pp. 669–80

21. Travaini, A., et al., 'Evaluation of neophobia and its potential impact upon predator control techniques: a study on two sympatric foxes in southern Patagonia', *Behavioural Processes*, 2013, 92, pp. 79–87

22. Misslin, R. and M. Cigrang, 'Does neophobia necessarily imply fear or anxiety?', *Behavioural Processes*, 1986, 12(1), pp. 45–50

23. Quintero, E., et al., 'Effects of context novelty vs. familiarity on latent inhibition with a conditioned taste aversion procedure', *Behavioural Processes*, 2011, 86(2), pp. 242–9

24. Brocklin, E. V., *The Science of Homesickness*, Duke Alumni, 2014

25. Bhugra, D. and M. A. Becker, 'Migration, cultural bereavement and

cultural identity', *World Psychiatry*, 2005, 4(1), pp. 18–24

26. Silove, D., P. Ventevogel and S. Rees, 'The contemporary refugee crisis: an overview of mental health challenges', *World Psychiatry*, 2017, 16(2), pp. 130–9

27. Holmes, T. and R. Rahe, 'The Holmes–Rahe life changes scale', *Journal of Psychosomatic Research*, 1967, 11, pp. 213–18

28. Zhang, R., T. J. Brennan and A. W. Lo, 'The origin of risk aversion', *Proceedings of the National Academy of Sciences*, 2014, 111(50), pp. 17777–82

29. Ickes, B. R., et al., 'Long-term environmental enrichment leads to regional increases in neurotrophin levels in rat brain', *Experimental Neurology*, 2000, 164(1), pp. 45–52

30. Young, D., et al., 'Environmental enrichment inhibits spontaneous apoptosis, prevents seizures and is neuroprotective', *Nature Medicine*, 1999, 5(4)

31. Hicklin, A., 'How Brooklyn became a writers' mecca', *Guardian*, 7 July 2012

32. Quintero, E., et al., 'Effects of context novelty vs. familiarity on latent inhibition with a conditioned taste aversion procedure', *Behavioural Processes*, 2011, 86(2), pp. 242–9

33. Bouter, L. M., et al., 'Sensation seeking and injury risk in downhill skiing', *Personality and Individual Differences*, 1988, 9(3), pp. 667–73

34. Smith, S. G., 'The essential qualities of a home', *Journal of Environmental Psychology*, 1994, 14(1), pp. 31–46

35. Hall, E. T., *The Hidden Dimension*, Doubleday, 1966

36. Aiello, J. R. and D. E. Thompson, 'Personal space, crowding, and spatial behavior in a cultural context', *Environment and Culture*, 1980, pp. 107–78

37. Lourenco, S. F., M. R. Longo and T. Pathman, 'Near space and its relation to claustrophobic fear', *Cognition*, 2011, 119(3), pp. 448–53

38. Kennedy, D. P., et al., 'Personal space regulation by the human amygdala', *Nature Neuroscience*, 2009, 12(10), pp. 1226–7

39. Evans, G. W. and R. E. Wener, 'Crowding and personal space invasion on the train: Please don't make me sit in the middle', *Journal of Environmental Psychology*, 2007, 27(1), pp. 90–94

40. Schwartz, B., 'The social psychology of privacy', *American Journal of Sociology*, 1968, pp. 741–52

41. Berman, M. G., J. Jonides and S. Kaplan, 'The cognitive benefits of interacting with nature', *Psychological Science*, 2008, 19(12), pp. 1207–12

42. Ulrich, R., 'View through a window may influence recovery', *Science*, 1984, 224(4647), pp. 224–5

43. Dobbs, D., 'The green space cure: the psychological value of

biodiversity', *Scientific American*, 13 November 2007

44. 'Tiny house movement', Wikipedia, 2017, wikipedia.org/wiki/Tiny_house_movement

45. Bouchard, T. J., 'Genes, environment, and personality', *Science*, 1994, p. 1700

46. Oishi, S. and U. Schimmack, 'Residential mobility, well-being, and mortality', *Journal of Personality and Social Psychology*, 2010, 98(6), p. 980

47. Jang, Y. and D. E. Huber, 'Context retrieval and context change in free recall: recalling from long-term memory drives list isolation', *Journal of Experimental Psychology: Learning, Memory, and Cognition*, 2008, 34(1), p. 112

48. Rubinstein, R. L., 'The home environments of older people: a description of the psychosocial processes linking person to place', *Journal of Gerontology*, 1989, 44(2), pp. S45–S53.

49. Winograd, E. and W. A. Killinger, 'Relating age at encoding in early childhood to adult recall: development of flashbulb memories', *Journal of Experimental Psychology: General*, 1983, 112(3), p. 413

50. Lollar, K., 'The liminal experience: loss of extended self after the fire', *Qualitative Inquiry*, 2009

51. Jones, R. T. and D. P. Ribbe, 'Child, adolescent, and adult victims of residential fire: psychosocial consequences', *Behavior Modification*, 1991, 15(4), pp. 560–80

52. Kim, K. and M. K. Johnson, 'Extended self: medial prefrontal activity during transient association of self and objects', *Social Cognitive and Affective Neuroscience*, 2010, pp. 199-207

53. Proshansky, H. M., A. K. Fabian and R. Kaminoff, 'Place-identity: physical world socialization of the self', *Journal of Environmental Psychology*, 1983, 3(1), pp. 57–83

54. Anton, C. E. and C. Lawrence, 'Home is where the heart is: the effect of place of residence on place attachment and community participation', *Journal of Environmental Psychology*, 2014, 40, pp. 451–61

Chapter 3: Working on the Brain

1. 'University of Bologna', Wikipedia, 2017, wikipedia.org/wiki/University_of_Bologna

2. Wilson, M., 'Stunning documentary looks at life inside a marble mine', *Fast Company*, 14 November 2014, fastcodesign.com

3. 'What Percentage of Your Life Will You Spend at Work? ',

ReviseSociology.com, 2016, @realsociology

4. Work-related Stress, Anxiety and Depression Statistics in Great Britain, Health and Safety Executive, 2016, hse.gov.uk/statistics/causdis/stress/

5. Number of Jobs, Labor Market Experience, and Earnings Growth: Results from a Longitudinal Survey, Bureau of Labor Statistics, 2017, bls.gov/news.release/nlsoy.toc.htm

6. Erickson, K. I., C. H. Hillman and A. F. Kramer, 'Physical activity, brain, and cognition', *Current Opinion in Behavioral Sciences*, 2015, 4(Supplement C), pp. 27–32

7. Swaminathan, N., 'Why does the brain need so much power?', *Scientific American*, 2008 29(04), p. 2998

8. Sleiman, S. F., et al., 'Exercise promotes the expression of brain derived neurotrophic factor (BDNF) through the action of the ketone body β-hydroxybutyrate', *Elife*, 2016, 5, p. e15092

9. Godman, H., 'Regular exercise changes the brain to improve memory, thinking skills', *Harvard Health Letters*, 2014

10. White, L. J. and V. Castellano, 'Exercise and brain health – implications for multiple sclerosis', *Sports Medicine*, 2008, 38(2), pp. 91–100

11. Kohl, H. W. and H. D. Cook, 'Physical activity, fitness, and physical education: effects on academic performance', in *Educating the Student Body: Taking Physical Activity and Physical Education to School*, National Academies Press, 2013

12. Gonzalez-Mulé, E., K. M. Carter and M. K. Mount, 'Are smarter people happier? Meta-analyses of the relationships between general mental ability and job and life satisfaction', *Journal of Vocational Behavior*, 2017, 99(Supplement C), pp. 146–64

13. Thorén, P., et al., 'Endorphins and exercise: physiological mechanisms and clinical implications', *Medicine and Science in Sports and Exercise*, 1990

14. Almeida, R. P., et al., 'Effect of cognitive reserve on age-related changes in cerebrospinal fluid biomarkers of Alzheimer disease', *JAMA Neurology*, 2015, 72(6), pp. 699–706

15. Scarmeas, N. and Y. Stern, 'Cognitive reserve: implications for diagnosis and prevention of Alzheimer's disease', *Current Neurology and Neuroscience Reports*, 2004, 4(5), pp. 374–380

16. Kurniawan, I. T., et al., 'Effort and valuation in the brain: the effects of anticipation and execution', *Journal of Neuroscience*, 2013, 33(14), p. 6160

17. Hagura, N., P. Haggard and J. Diedrichsen, 'Perceptual decisions are biased by the cost to act', *Elife*, 2017, 6, p. e18422

18. Herz, R S. and J. von Clef, 'The influence of verbal labeling on the

perception of odors: evidence for olfactory illusions?', *Perception*, 2001, 30(3), pp. 381–91

19. Elliott, R., et al., 'Differential response patterns in the striatum and orbitofrontal cortex to financial reward in humans: a parametric functional magnetic resonance imaging study', *Journal of Neuroscience*, 2003, 23(1), p. 303

20. Holmes, T. and R. Rahe, 'Holmes–Rahe life changes scale', *Journal of Psychosomatic Research*, 1967, 11, pp. 213–18

21. Howell, R. T., M. Kurai and L. Tam, 'Money buys financial security and psychological need satisfaction: testing need theory in affluence', *Social Indicators Research*, 2013, 110(1), pp. 17–29

22. Sheldon, K. M. and A. Gunz, 'Psychological needs as basic motives, not just experiential requirements', *Journal of Personality*, 2009, 77(5), pp. 1467–92

23. Roddenberry, A. and K. Renk, 'Locus of control and self-efficacy: potential mediators of stress, illness, and utilization of health services in college students', *Child Psychiatry and Human Development*, 2010, 41(4), pp. 353–370

24. Abramowitz, S. I., 'Locus of control and self-reported depression among college students', *Psychological Reports*, 1969, 25(1), pp. 149–150

25. Williams, J. S., et al., 'Health locus of control and cardiovascular risk factors in veterans with Type 2 diabetes', *Endocrine*, 2016, 51(1), pp. 83–90

26. Lefcourt, H. M., *Locus of Control: Current Trends in Theory and Research*, Psychology Press, 2014

27. Pruessner, J. C., et al., 'Self-esteem, locus of control, hippocampal volume, and cortisol regulation in young and old adulthood', *NeuroImage*, 2005, 28(4), pp. 815–26

28. Lewis, M., S. M. Alessandri and M. W. Sullivan, 'Violation of expectancy, loss of control, and anger expressions in young infants', *Developmental Psychology*, 1990, 26(5), p. 745

29. Leavitt, L. A. and W. L. Donovan, 'Perceived infant temperament, locus of control, and maternal physiological response to infant gaze', *Journal of Research in Personality*, 1979, 13(3), pp. 267–78

30. Colles, S. L., J. B. Dixon and P. E. O'Brien, 'Loss of control is central to psychological disturbance associated with binge eating disorder', *Obesity*, 2008, 16(3), pp. 608–14

31. Rosen, H. J., et al., 'Neuroanatomical correlates of cognitive self-appraisal in neurodegenerative disease', *NeuroImage*, 2010, 49(4), pp. 3358–64

32. Maguire, E. A., K. Woollett and H. J. Spiers, 'London taxi drivers and bus drivers: a structural MRI and neuropsychological analysis',

Hippocampus, 2006, 16(12), pp. 1091–1101

33. Gaser, C. and G. Schlaug, 'Brain structures differ between musicians and non-musicians', *Journal of Neuroscience*, 2003, 23(27), pp. 9240–5

34. Castelli, F., D. E. Glaser and B. Butterworth, 'Discrete and analogue quantity processing in the parietal lobe: a functional MRI study', *Proceedings of the National Academy of Sciences of the United States of America*, 2006, 103(12), pp. 4693–8

35. Grefkes, C. and G. R. Fink, 'The functional organization of the intraparietal sulcus in humans and monkeys', *Journal of Anatomy*, 2005, 207(1), pp. 3–17

36. Oswald, A. J., E. Proto and D. Sgroi, 'Happiness and productivity', *Journal of Labor Economics*, 2015, 33(4), pp. 789–822

37. Farhud, D. D., M. Malmir and M. Khanahmadi, 'Happiness and health: the biological factors – systematic review article', *Iranian Journal of Public Health*, 2014, 43(11), p. 1468

38. Zwosta, K., H. Ruge and U. Wolfensteller, 'Neural mechanisms of goal-directed behavior: outcome-based response selection is associated with increased functional coupling of the angular gyrus', *Frontiers in Human Neuroscience*, 2015, 9

39. Elliot, A. J. and M. V. Covington, 'Approach and avoidance motivation', *Educational Psychology Review*, 2001, 13(2), pp. 73–92

40. Cofer, C. N., 'The history of the concept of motivation', *Journal of the History of the Behavioral Sciences*, 1981, 17(1), pp. 48–53

41. Lee, W., et al., 'Neural differences between intrinsic reasons for doing versus extrinsic reasons for doing: an fMRI study', *Neuroscience Research*, 2012, 73(1), pp. 68–72

42. Benabou, R. and J. Tirole, 'Intrinsic and extrinsic motivation', *Review of Economic Studies*, 2003, 70(3), pp. 489–520

43. Lepper, M. R., D. Greene and R. E. Nisbett, 'Undermining children's intrinsic interest with extrinsic reward: a test of the "overjustification" hypothesis', *Journal of Personality and Social Psychology*, 1973, 28(1), pp. 129–37

44. Lapierre, S., L. Bouffard and E. Bastin, 'Personal goals and subjective well-being in later life', *International Journal of Aging and Human Development*, 1997, 45(4), p. 287–303

45. Agnew, R., 'Foundation for a general strain theory of crime and delinquency', *Criminology*, 1992, 30(1), pp. 47–88

46. Higgins, E. T., et al., 'Ideal versus ought predilections for approach and avoidance distinct self-regulatory systems', *Journal of Personality and Social Psychology*, 1994, 66(2), p. 276

47. Leonard, N. H., L. L. Beauvais and R. W. Scholl, 'Work motivation: the incorporation of self-concept-based processes', *Human*

*Relations*, 1999, 52(8), pp. 969–98

48. Neal, D. T., W. Wood and A. Drolet, 'How do people adhere to goals when willpower is low? The profits (and pitfalls) of strong habits', *Journal of Personality and Social Psychology*, 2013, 104(6), p. 959

49. Bem, D. J., 'Self-perception: an alternative interpretation of cognitive dissonance phenomena', *Psychological Review*, 1967, 74(3), p. 183

50. Utevsky, A. V. and M. L. Platt, 'Status and the brain', *PLOS Biology*, 2014, 12(9), p. e1001941

51. Pezzulo, G., et al., 'The principles of goal-directed decision-making: from neural mechanisms to computation and robotics', *Philosophical Transactions of the Royal Society B*, 369(1655), 2014

52. Leung, B. K. and B. W. Balleine, 'The ventral striato-pallidal pathway mediates the effect of predictive learning on choice between goal-directed actions', *Journal of Neuroscience*, 2013, 33(34), p. 13848

53. Media, O., Nuffield Farming Scholarships Trust, 2017, nuffieldscholar.org

54. Miron-Shatz, T., '"Am I going to be happy and financially stable?" How American women feel when they think about financial security', *Judgment and Decision Making*, 2009, 4(1), pp. 102–112

55. Moesgaard, S. 'How money affects the brain's reward system (why money is addictive) ', reflectd.co, 21 March 2013

56. Hyman, S. E. and R. C. Malenka, 'Addiction and the brain: the neurobiology of compulsion and its persistence', *Nature Reviews Neuroscience*, 2001, 2(10), p. 695

57. Sharot, T., *The Optimism Bias: A Tour of the Irrationally Positive Brain*, Vintage, 2011

58. Howell, et al., 'Money buys financial security and psychological need satisfaction: testing need theory in affluence', *Social Indicators Research*, 2012

59. Holmes, T. and R. Rahe, 'The Holmes–Rahe life changes scale', *Journal of Psychosomatic Research*, 1967, 11, pp. 213–18

60. Saarni, C., *The Development of Emotional Competence*, Guilford Press, 1999

61. Rodriguez, T., 'Negative emotions are key to well-being', *Scientific American*, 1 May 2013

62. Adkins, A., 'U.S. employee engagement steady in June', 2016, GALLUP

63. Spicer, A. and C. Cederström, 'The research we've ignored about happiness at work', *Harvard Business Review*, 21 July 2015

64. Van Kleef, G. A., C. K. De Dreu and A. S. Manstead, 'The interpersonal effects of anger and happiness in negotiations', *Journal*

*of Personality and Social Psychology*, 2004, 86(1), pp. 57–76

65. Ferguson, D., 'The world's happiest jobs', *Guardian*, 7 April 2015

66. Peralta, C. F. and M. F. Saldanha, 'Can dealing with emotional exhaustion lead to enhanced happiness? The roles of planning and social support', *Work and Stress*, 2017, 31(2), pp. 121–44

67. Mauss, I. B., et al., 'The pursuit of happiness can be lonely', *Emotion*, 2012, 12(5), p. 908

## Chapter 4: Happiness Is Other People

1. Theeuwes, J., 'Top-down and bottom-up control of visual selection', *Acta Psychologica*, 2010, 135(2), pp. 77–99

2. LoBue, V., et al., 'What accounts for the rapid detection of threat? Evidence for an advantage in perceptual and behavioral responding from eye movements', *Emotion*, 2014, 14(4), pp. 816–23

3. Jabbi, M., J. Bastiaansen and C. Keysers, 'A common anterior insula representation of disgust observation, experience and imagination shows divergent functional connectivity pathways', *PLOS ONE*, 2008, 3(8), p. e2939

4. Clarke, D., 'Circulation and energy metabolism of the brain', *Basic Neurochemistry: Molecular, Cellular and Medical Aspects*, 1999, pp. 637–69

5. Miller, G., *The Mating Mind: How Sexual Choice Shaped the Evolution of Human Nature*, Anchor, 2011

6. Dunbar, R. I., 'The social brain hypothesis and its implications for social evolution', *Annals of Human Biology*, 2009, 36(5), pp. 562–72

7. Flinn, M. V., D. C. Geary and C. V. Ward, 'Ecological dominance, social competition, and coalitionary arms races: why humans evolved extraordinary intelligence', *Evolution and Human Behavior*, 2005, 26(1), pp. 10–46

8. Reader, S. M. and K. N. Laland, 'Social intelligence, innovation, and enhanced brain size in primates', *Proceedings of the National Academy of Sciences of the United States of America*, 2002, 99(7), pp. 4436–41

9. Spradbery, J. P., *Wasps: An Account of the Biology and Natural History of Social and Solitary Wasps*, Sidgwick & Jackson, 1973

10. Gavrilets, S., 'Human origins and the transition from promiscuity to pair-bonding', *Proceedings of the National Academy of Sciences of the United States of America*, 2012, 109(25), pp. 9923–8

11. West, R. J., 'The evolution of large brain size in birds is related to social, not genetic, monogamy', *Biological Journal of the Linnean Society*, 2014, 111(3), pp. 668–78

12. Bales, K. L., et al., 'Neural correlates of pair-bonding in a

monogamous primate', *Brain Research*, 2007, 1184, pp. 245–53

13. Dunbar, R. I. M. and S. Shultz, 'Evolution in the social brain', *Science*, 2007, 317(5843), pp. 1344–7

14. Pasquaretta, C., et al., 'Social networks in primates: smart and tolerant species have more efficient networks', *Scientific Reports*, 2014, 4, p. 7600

15. Van Gestel, S. and C. Van Broeckhoven, 'Genetics of personality: are we making progress?' *Molecular Psychiatry*, 2003, 8(10), pp. 840–52

16. Matsuzawa, T., 'Evolution of the brain and social behavior in chimpanzees', *Current Opinion in Neurobiology*, 2013, 23(3), pp. 443–9

17. Gunaydin, Lisa A., et al., 'Natural neural projection dynamics underlying social behavior', *Cell*, 157(7), pp. 1535–51

18. Gardner, E. L., 'Introduction: addiction and brain reward and anti-reward pathways', *Advances in Psychosomatic Medicine*, 2011, 30, pp. 22–60

19. Loken, L. S., et al., 'Coding of pleasant touch by unmyelinated afferents in humans', *Nature Neuroscience*, 2009, 12(5), pp. 547–8

20. Iggo, A., 'Cutaneous mechanoreceptors with afferent C fibres', *Journal of Physiology*, 1960, 152(2), pp. 337–53

21. 'Insular cortex', Wikipedia, 2017, wikipedia.org/wiki/Insular_cortex

22. Kalueff, A. V., J. L. La Porte and C. L. Bergner, *Neurobiology of Grooming Behavior*, Cambridge University Press, 2010

23. Claxton, G., 'Why can't we tickle ourselves?', *Perceptual and Motor Skills*, 1975, 41(1), pp. 335–8

24. Keverne, E. B., N. D. Martensz and B. Tuite, 'Beta-endorphin concentrations in cerebrospinal fluid of monkeys are influenced by grooming relationships', *Psychoneuroendocrinology*, 1989, 14(1), pp. 155–61

25. Gispen, W. H., et al., 'Modulation of ACTH-induced grooming by [DES-TYR1]-γ-endorphin and haloperidol', *European Journal of Pharmacology*, 1980, 63(2), pp. 203–7

26. Dumbar, R., 'Co-evolution of neocortex size, group size and language in humans', *Behavioral and Brain Sciences*, 1993, 16(4), pp. 681–735

27. Dunbar, R. and R. I. M. Dunbar, *Grooming, Gossip, and the Evolution of Language*, Harvard University Press, 1998

28. Crusco, A. H. and C. G. Wetzel, 'The Midas touch', *Personality and Social Psychology Bulletin*, 1984, 10(4), pp. 512–17

29. Dumas, G., et al., 'Inter-brain synchronization during social interaction', *PLOS ONE*, 2010, 5(8), p. e12166

30. Livingstone, M. S. and D. H. Hubel, 'Anatomy and physiology of a color system in the primate visual cortex', *Journal of Neuroscience*,

1984, 4(1), pp. 309–56

31. Rizzolatti, G., et al., 'From mirror neurons to imitation: facts and speculations', *The Imitative Mind: Development, Evolution, and Brain Bases*, 2002, 6, pp. 247–66

32. Wicker, B., et al., 'Both of us disgusted in my insula', *Neuron*, 2003, 40(3), pp. 655–64

33. Schulte-Rüther, M., et al., 'Mirror neuron and theory of mind mechanisms involved in face-to-face interactions: a functional magnetic resonance imaging approach to empathy', *Journal of Cognitive Neuroscience*, 2007, 19(8), pp. 1354–72

34. Shamay-Tsoory, S. G., J. Aharon-Peretz and D. Perry, 'Two systems for empathy: a double dissociation between emotional and cognitive empathy in inferior frontal gyrus versus ventromedial prefrontal lesions', *Brain*, 2009, 132(3), pp. 617–27

35. de Waal, F. B. M., 'Apes know what others believe', *Science*, 2016, 354(6308), p. 39

36. Brink, T. T., et al., 'The role of orbitofrontal cortex in processing empathy stories in four- to eight-year-old children', *Frontiers in Psychology*, 2011, 2, p. 80

37. Hall, F. S., 'Social deprivation of neonatal, adolescent, and adult rats has distinct neurochemical and behavioral consequences', *Critical Reviews in Neurobiology*, 1998, 12(1–2)

38. Martin, L. J., et al., 'Social deprivation of infant rhesus monkeys alters the chemoarchitecture of the brain: I. Subcortical regions', *Journal of Neuroscience*, 1991, 11(11), pp. 3344–58

39. Metzner, J. L. and J. Fellner, 'Solitary confinement and mental illness in US prisons: a challenge for medical ethics', *Journal of the American Academy of Psychiatry and the Law*, 2010, 38(1), pp. 104–8

40. Izuma, K., D. N. Saito and N. Sadato, 'Processing of the incentive for social approval in the ventral striatum during charitable donation', *Journal of Cognitive Neuroscience*, 2010, 22(4), pp. 621–31

41. Buchanan, K. E. and A. Bardi, 'Acts of kindness and acts of novelty affect life satisfaction', *Journal of Social Psychology*, 2010, 150(3), pp. 235–7

42. Bateson, M., D. Nettle and G. Roberts, 'Cues of being watched enhance cooperation in a real-world setting', *Biology Letters*, 2006, 2(3), pp. 412–14

43. Rigdon, M., et al., 'Minimal social cues in the dictator game', *Journal of Economic Psychology*, 2009, 30(3), pp. 358–67

44. Weir, K., 'The pain of social rejection', *American Psychological Association*, 2012, 43

45. Woo, C. W., et al., 'Separate neural representations for physical pain

and social rejection', *Nature Communications*, 2014, 5, p. 5380

46. Wesselmann, E. D., et al., 'Adding injury to insult: unexpected rejection leads to more aggressive responses', *Aggressive Behavior*, 2010, 36(4), pp. 232–7

47. Farrow, T., et al., 'Neural correlates of self-deception and impression-management', *Neuropsychologia*, 2014, 67

48. Morrison, S., J. Decety and P. Molenberghs, 'The neuroscience of group membership', *Neuropsychologia*, 2012, 50(8), pp. 2114–20

49. D'Argembeau, A., 'On the role of the ventromedial prefrontal cortex in self-processing: the valuation hypothesis', *Frontiers in Human Neuroscience*, 2013, 7, p. 372

50. Fischer, P., et al., 'The bystander-effect: a meta-analytic review on bystander intervention in dangerous and non-dangerous emergencies', *Psychological Bulletin*, 2011, 137(4), p. 517

51. Gonçalves, B., N. Perra and A. Vespignani, 'Modeling users' activity on Twitter networks: validation of Dunbar's number', *PLOS ONE*, 2011, 6(8), p. e22656

## Chapter 5: Love, Lust or Bust

1. Clark, C., 'Brain sex in men and women – from arousal to orgasm', *BrainBlogger*, 2014

2. Laeng, B., O. Vermeer and U. Sulutvedt, 'Is beauty in the face of the beholder?', *PLOS ONE*, 2013, 8(7), p. e68395

3. Järvi, T., et al., 'Evolution of variation in male secondary sexual characteristics', *Behavioral Ecology and Sociobiology*, 1987, 20(3), pp. 161–9

4. Georgiadis, J. R. and M. L. Kringelbach, 'Intimacy and the brain: lessons from genital and sexual touch', in Olausson, H., et al. (eds), *Affective Touch and the Neurophysiology of CT Afferents*, Springer, 2016, pp. 301–21

5. Cazala, F., N. Vienney and S. Stoléru, 'The cortical sensory representation of genitalia in women and men: a systematic review', *Socioaffective Neuroscience and Psychology*, 2015, 5, p. 10.3402/snp. v5.26428.

6. 'The neuroscience of erogenous zones', 2017, www.bangor.ac.uk/ psychology/news/the-neuroscience-of-erogenous-zones-15794.

7. Turnbull, O. H., et al., 'Reports of intimate touch: Erogenous zones and somatosensory cortical organization', *Cortex*, 2014, 53, pp. 146–54

8. Georgiadis, J. R., 'Doing it . . . wild? On the role of the cerebral cortex in human sexual activity', *Socioaffective Neuroscience and*

*Psychology*, 2012, 2, p. 17337

9. Aggleton, E. J. P., et al., *The Amygdala: A Functional Analysis*, Oxford University Press, 2000

10. Baird, A. D., et al., 'The amygdala and sexual drive: insights from temporal lobe epilepsy surgery', *Annals of Neurology*, 2004, 55(1), pp. 87–96

11. Newman, S. W., 'The medial extended amygdala in male reproductive behavior: a node in the mammalian social behavior network', *Annals of the New York Academy of Sciences*, 1999, 877(1), pp. 242–57

12. Goldstein, J. M., 'Sex, hormones and affective arousal circuitry dysfunction in schizophrenia', *Hormones and Behavior*, 2006, 50(4), pp. 612–22

13. Shirtcliff, E. A., R. E. Dahl and S. D. Pollak, 'Pubertal development: correspondence between hormonal and physical development', *Child Development*, 2009, 80(2), pp. 327–37

14. Alexander, G. M. and B. B. Sherwin, 'The association between testosterone, sexual arousal, and selective attention for erotic stimuli in men', *Hormones and Behavior*, 1991, 25(3), pp. 367–81

15. van Anders, S. M., 'Testosterone and sexual desire in healthy women and men', *Archives of Sexual Behavior*, 2012, 41(6), pp. 1471–84

16. Rajfer, J., 'Relationship between testosterone and erectile dysfunction', *Reviews in Urology*, 2000, 2(2), pp. 122–8

17. Sarrel, P. M., 'Effects of hormone replacement therapy on sexual psychophysiology and behavior in postmenopause', *Journal of Women's Health and Gender-Based Medicine*, 2000, 9(1, Supplement 1), pp. 25–32

18. Sarrel, P., B. Dobay and B. Wiita, 'Estrogen and estrogen-androgen replacement in postmenopausal women dissatisfied with estrogen-only therapy: sexual behavior and neuroendocrine responses', *Journal of Reproductive Medicine*, 1998, 43(10), pp. 847–56

19. Purves, D., G. Augustine and D. Fitzpatrick, 'Autonomic regulation of sexual function', *Neuroscience*, Sinauer Associates, 2001

20. Ishai, A., 'Sex, beauty and the orbitofrontal cortex', *International Journal of Psychophysiology*, 2007, 63(2), pp. 181–5

21. Ortega, V., I. Zubeidat and J. C. Sierra, 'Further examination of measurement properties of Spanish version of the Sexual Desire Inventory with undergraduates and adolescent students', *Psychological Reports*, 2006, 99(1), pp. 147–65

22. Montgomery, K. A., 'Sexual desire disorders', *Psychiatry*, 2008, 5(6), pp. 50–55

23. Gray, J. A., 'Brain systems that mediate both emotion and cognition', *Cognition and Emotion*, 1990, 4(3), pp. 269–88

24. Swerdlow, N. R. and G. F. Koob, 'Dopamine, schizophrenia, mania, and depression: toward a unified hypothesis of cortico-striatopallido-thalamic function', *Behavioral and Brain Sciences*, 1987, 10(2), pp. 197–208

25. Shenhav, A., M. M. Botvinick and J. D. Cohen, 'The expected value of control: an integrative theory of anterior cingulate cortex function', *Neuron*, 2013, 79(2), pp. 217–40

26. Gola, M., M. Miyakoshi and G. Sescousse, 'Sex, impulsivity, and anxiety: interplay between ventral striatum and amygdala reactivity in sexual behaviors', *Journal of Neuroscience*, 2015, 35(46), p. 15227

27. McCabe, M. P., 'The role of performance anxiety in the development and maintenance of sexual dysfunction in men and women', *International Journal of Stress Management*, 2005, 12(4), pp. 379–88

28. Welborn, B. L., et al., 'Variation in orbitofrontal cortex volume: relation to sex, emotion regulation and affect', *Social Cognitive and Affective Neuroscience*, 2009, 4(4), pp. 328–39

29. Spinella, M., 'Clinical case report: hypersexuality and dysexecutive syndrome after a thalamic infarct', *International Journal of Neuroscience*, 2004, 114(12), pp. 1581–90

30. Stoléru, S., et al., 'Brain processing of visual sexual stimuli in men with hypoactive sexual desire disorder', *Psychiatry Research: Neuroimaging*, 2003, 124(2), pp. 67–86

31. Freeman, S. 'What happens in the brain during an orgasm?', 2008, health.howstuffworks.com/sexual-health/sexuality/brain-during-orgasm.htm

32. Pfaus, J. G., 'Reviews: pathways of sexual desire', *Journal of Sexual Medicine*, 2009, 6(6), pp. 1506–33

33. Georgiadis, J. R., et al., 'Men versus women on sexual brain function: prominent differences during tactile genital stimulation, but not during orgasm', *Human Brain Mapping*, 2009, 30(10), pp. 3089–3101

34. Komisaruk, B. R. and B. Whipple, 'Functional MRI of the brain during orgasm in women', *Annual Review of Sex Research*, 2005, 16(1), pp. 62–86

35. Komisaruk, B., et al. 'An fMRI time-course analysis of brain regions activated during self stimulation to orgasm in women', *Society for Neuroscience Abstracts*, 2010

36. Hunter, A., 'Orgasm just by thinking: is it medically possible?', 19 July 2010, cbsnews.com

37. Park, B. Y., et al., 'Is internet pornography causing sexual dysfunctions? A review with clinical reports', *Behavioral Sciences*,

2016, 6(3), p. 17

38. Opie, C., et al., 'Male infanticide leads to social monogamy in primates', *Proceedings of the National Academy of Sciences*, 2013, 110(33), pp. 13328–32

39. Comninos, A. N., et al., 'Kisspeptin modulates sexual and emotional brain processing in humans', *Journal of Clinical Investigation*, 2017, 127(2), p. 709

40. Cho, M. M., et al., 'The effects of oxytocin and vasopressin on partner preferences in male and female prairie voles (Microtus ochrogaster)', 1999, *Behavioral Neuroscience*, 113(5), pp. 1071–9

41. Gardner, E. L., 'Introduction: addiction and brain reward and anti-reward pathways', *Advances in Psychosomatic Medicine*, 2011, 30, pp. 22–60

42. Nephew, B. C., 'Behavioral roles of oxytocin and vasopressin', in T. Sumiyoshi (ed.), *Neuroendocrinology and Behavior*, InTech, 2012

43. Bales, K. L., et al., 'Neural correlates of pair-bonding in a monogamous primate', *Brain Research*, 2007, 1184, pp. 245–53

44. Young, L. J. and Z. Wang, 'The neurobiology of pair bonding', *Nature Neuroscience*, 2004, 7(10), pp. 1048–54

45. Lim, M. M., et al., 'Enhanced partner preference in a promiscuous species by manipulating the expression of a single gene', *Nature*, 2004, 429(6993), p. 754

46. Lim, M. M., E. A. D. Hammock and L. J. Young, 'The role of vasopressin in the genetic and neural regulation of monogamy', *Journal of Neuroendocrinology*, 2004, 16(4), pp. 325–32

47. Fisher, H. E., et al., 'Defining the brain systems of lust, romantic attraction, and attachment', *Archives of Sexual Behavior*, 2002, 31(5), pp. 413–19

48. Brown, N. J., A. D. Sokal and H. L. Friedman, 'The complex dynamics of wishful thinking: the critical positivity ratio', *American Psychologist*, 2013, 68(9), pp. 801–13

49. Kottemann, K. L., 'The rhetoric of deliberate deception: what catfishing can teach us', University of Louisiana at Lafayette, 2015

50. Aron, A., et al., 'Reward, motivation, and emotion systems associated with early-stage intense romantic love', *Journal of Neurophysiology*, 2005, 94(1), pp. 327–37

51. Fisher, H., 'The drive to love: the neural mechanism for mate selection', *New Psychology of Love*, 2006, pp. 87–115

52. Savulescu, J. and A. Sandberg, 'Neuroenhancement of love and marriage: the chemicals between us', *Neuroethics*, 2008, 1(1), pp. 31–44

53. Dayan, P. and Q. J. Huys, 'Serotonin, inhibition, and negative mood', *PLOS Computational Biology*, 2008, 4(2), p. e4

54. Portas, C. M., B. Bjorvatn and R. Ursin, 'Serotonin and the sleep/ wake cycle: special emphasis on microdialysis studies', *Progress in Neurobiology*, 2000, 60(1), pp. 13–35

55. Hesse, S., et al., 'Serotonin and dopamine transporter imaging in patients with obsessive-compulsive disorder', *Psychiatry Research: Neuroimaging*, 2005, 140(1), pp. 63–72

56. Wood, H., 'Love on the brain', *Nature Reviews Neuroscience*, 2001, 2(2), p. 80

57. Zeki, S., 'The neurobiology of love', *FEBS Letters*, 2007, 581(14), pp. 2575–9

58. Johnson-Laird, P. N., 'Mental models and human reasoning', *Proceedings of the National Academy of Sciences*, 2010, 107(43), pp. 18243–50

59. Acevedo, B. P., et al., 'Neural correlates of long-term intense romantic love', *Social Cognitive and Affective Neuroscience*, 2012, 7(2), pp. 145–59

60. Boynton, P. M., *The Research Companion: A Practical Guide for Those in Social Science, Health and Development*, Taylor and Francis, 2016

61. 'Arranged/forced marriage statistics', *Statistic Brain*, 2016, statisticbrain.com/arranged-marriage-statistics/

62. Gahran, A., *Stepping Off the Relationship Escalator: Uncommon Love and Life*, Off the Escalator Enterprises, 2017

63. Twenge, J. M., R. A. Sherman and B. E. Wells, 'Changes in American adults' reported same-sex sexual experiences and attitudes, 1973–2014', *Archives of Sexual Behavior*, 2016, 45(7), pp. 1713–30

64. Girl on the Net, 'Sexy stories, mostly true', 2017, girlonthenet.com

65. Girl on the Net, *Girl on the Net: How a Bad Girl Fell in Love*, BLINK Publishing, 2016

66. Wilson, G. D., 'Male–female differences in sexual activity, enjoyment and fantasies', *Personality and Individual Differences*, 1987, 8(1), pp. 125–7

67. Levin, R. and A. Riley, 'The physiology of human sexual function', *Psychiatry*, 2007, 6(3), pp. 90–94

68. McQuaid, J., 'Why we love the pain of spicy food', *Wall Street Journal*, 31 December 2014

69. Person, E. S., 'Sexuality as the mainstay of identity: psychoanalytic perspectives', *Signs: Journal of Women in Culture and Society*, 1980, 5(4), pp. 605–630

70. Weaver, H., G. Smith and S. Kippax, 'School-based sex education policies and indicators of sexual health among young people: a comparison of the Netherlands, France, Australia and the United

States', *Sex Education*, 2005, 5(2), pp. 171–88

71. Potard, C., et al., 'The relationship between parental attachment and sexuality in early adolescence', *International Journal of Adolescence and Youth*, 2017, 22(1), pp. 47–56

72. Hoffmann, H., E. Janssen and S. L. Turner, 'Classical conditioning of sexual arousal in women and men: effects of varying awareness and biological relevance of the conditioned stimulus', *Archives of Sexual Behavior*, 2004, 33(1), pp. 43–53

73. Hatzenbuehler, M. L., J. C. Phelan and B. G. Link, 'Stigma as a fundamental cause of population health inequalities', *American Journal of Public Health*, 2013, 103(5), pp. 813–21

## Chapter 6: You've Got to Laugh

1. Winston, J. S., J. O'Doherty and R. J. Dolan, 'Common and distinct neural responses during direct and incidental processing of multiple facial emotions', *NeuroImage*, 2003, 20(1), pp. 84–97

2. Davila-Ross, M., et al., 'Chimpanzees (pan troglodytes) produce the same types of "laugh Faces" when they emit laughter and when they are silent', *PLOS ONE*, 2015, 10(6), p. e0127337

3. Ross, M. D., M. J. Owren and E. Zimmermann, 'Reconstructing the evolution of laughter in great apes and humans', *Current Biology*, 2009, 19(13), pp. 1106–11

4. Panksepp, J. and J. Burgdorf, '50-kHz chirping (laughter?) in response to conditioned and unconditioned tickle-induced reward in rats: effects of social housing and genetic variables', *Behavioural Brain Research*, 2000, 115(1), pp. 25–38

5. Weisfeld, G. E., 'The adaptive value of humor and laughter', *Ethology and Sociobiology*, 1993, 14(2), pp. 141–69

6. Pellis, S. and V. Pellis, *The Playful Brain: Venturing to the Limits of Neuroscience*, Oneworld Publications, 2013

7. Wild, B., et al., 'Neural correlates of laughter and humour', *Brain*, 2003, 126(10), pp. 2121–38

8. Selden, S. T., 'Tickle', *Journal of the American Academy of Dermatology*, 2004, 50(1), pp. 93–7

9. Claxton, G., 'Why can't we tickle ourselves?', *Perceptual and Motor Skills*, 1975, 41(1), pp. 335–8

10. Berman, R., 'The psychology of tickling and why it makes us laugh', *Big Think*, 2016, bigthink.com

11. Stafford, T., 'Why all babies love peekaboo', *BBC Future*, 2014, bbc.com

12. Vrticka, P., J. M. Black and A. L. Reiss, 'The neural basis of humour

processing', *Nature Reviews Neuroscience*, 2013, 14(12), pp. 860–8

13. Messinger, D. S., A. Fogel and K. L. Dickson, 'All smiles are positive, but some smiles are more positive than others', *Developmental Psychology*, 2001, 37(5), pp. 642–53

14. Scott, S., 'Beyond a joke: how to study laughter', *Guardian*, 10 July 2014

15. Chan, Y. C., et al., 'Towards a neural circuit model of verbal humor processing: an fMRI study of the neural substrates of incongruity detection and resolution', *NeuroImage*, 2013, 66, pp. 169–76

16. Hempelmann, C. F. and S. Attardo, 'Resolutions and their incongruities: further thoughts on logical mechanisms', *Humor*, 2011, 24(2), pp. 125–49

17. Franklin, R. G. Jr and R. B. Adams Jr, 'The reward of a good joke: neural correlates of viewing dynamic displays of stand-up comedy', *Cognitive, Affective and Behavioral Neuroscience*, 2011, 11(4), pp. 508–15

18. Pessoa, L. and R. Adolphs, 'Emotion processing and the amygdala: from a "low road" to "many roads" of evaluating biological significance', *Nature Reviews Neuroscience*, 2010, 11(11), p. 773

19. Scott, S. K., et al., 'The social life of laughter', *Trends in Cognitive Sciences*, 2014, 18(12), pp. 618–20

20. Prof Sophie Scott, 2017, ucl.ac.uk/pals/people/profiles/academic-staff/sophie-scott

21. Berk, L. S., et al., 'Neuroendocrine and stress hormone changes during mirthful laughter', *American Journal of the Medical Sciences*, 1989, 298(6), pp. 390–6

22. Dunbar, R. I., et al., 'Social laughter is correlated with an elevated pain threshold', *Proceedings of the Royal Society B: Biological Sciences*, 2012, 279(1731), pp. 1161–7

23. Manninen, S., et al., 'Social laughter triggers endogenous opioid release in humans', *Journal of Neuroscience*, 2017, 37(25), p. 6125

24. Wildgruber, D., et al., 'Different types of laughter modulate connectivity within distinct parts of the laughter perception network', *PLOS ONE*, 2013, 8(5), p. e63441

25. Philippon, A. C., L. M. Randall and J. Cherryman, 'The impact of laughter in earwitness identification performance', *Psychiatry, Psychology and Law*, 2013, 20(6), pp. 887–98

26. Uekermann, J., et al., 'Theory of mind, humour processing and executive functioning in alcoholism', *Addiction*, 2007, 102(2), pp. 232–40

27. Samson, A. C., et al., 'Perception of other people's mental states affects humor in social anxiety', *Journal of Behavior Therapy and Experimental Psychiatry*, 2012, 43(1), pp. 625–31

28. Wu, C.-L., et al., 'Do individuals with autism lack a sense of humor? A study of humor comprehension, appreciation, and styles among high school students with autism', *Research in Autism Spectrum Disorders*, 2014, 8(10), pp. 1386–93

29. Raine, J., 'The evolutionary origins of laughter are rooted more in survival than enjoyment', *The Conversation*, 13 April 2016

30. Gervais, M. and D. S. Wilson, 'The evolution and functions of laughter and humor: a synthetic approach', *Quarterly Review of Biology*, 2005, 80(4), pp. 395–430

31. Goldstein, J. H. 'Cross cultural research: humour here and there', in A. J. Chapman and H. C. Foot (eds), *It's a Funny Thing, Humor*, Elsevier, 1977

32. Provine, R. R. and K. Emmorey, 'Laughter among deaf signers', *Journal of Deaf Studies and Deaf Education*, 2006, 11(4), pp. 403–9

33. Davila-Ross, M., et al., 'Chimpanzees (pan troglodytes) produce the same type of "laugh faces" when they emit laughter and when they are silent', *PLOS ONE*, 2015, 10(6), p. e0127337

34. Cowan, M. L. and A. C. Little, 'The effects of relationship context and modality on ratings of funniness', *Personality and Individual Differences*, 2013, 54(4), pp. 496–500

35. Benazzi, F. and H. Akiskal, 'Irritable-hostile depression: further validation as a bipolar depressive mixed state', *Journal of Affective Disorders*, 2005, 84(2), pp. 197–207

36. WalesOnline, 'No joking but comedian Rhod is Wales' sexiest man', 2010, walesonline.co.uk/lifestyle/showbiz/no-joking-comedian-rhod-wales-1878454

37. Krebs, R., et al., 'Novelty increases the mesolimbic functional connectivity of the substantia nigra/ventral tegmental area (SN/VTA) during reward anticipation: evidence from high-resolution fMRI', *NeuroImage*, 2011, 58(2), pp. 647–55

38. Boldsworth, I., *The Mental Podcast*, 2017, ianboldsworth.co.uk/the-mental-podcast/

39. Boldsworth, I., *The ParaPod*, 2017, ianboldsworth.co.uk/project/the-parapod/

40. Hyman, S. E. and R. C. Malenka, 'Addiction and the brain: the neurobiology of compulsion and its persistence', *Nature Reviews Neuroscience*, 2001, 2(10), p. 695

41. Heimberg, R. G., *Social Phobia: Diagnosis, Assessment, and Treatment,* Guilford Press, 1995

42. Atkinson, J. W., 'Motivational determinants of risk-taking behavior', *Psychological Review*, 1957, 64(6 pt 1), p. 359

43. Samson, A. C. and J. J. Gross, 'Humour as emotion regulation: the differential consequences of negative versus positive humour',

*Cognition and Emotion*, 2012, 26(2), pp. 375–84

44. Gil, M., et al., 'Social reward: interactions with social status, social communication, aggression, and associated neural activation in the ventral tegmental area', *European Journal of Neuroscience*, 2013, 38(2), pp. 2308–18

45. Goh, C. and M. Agius, 'The stress-vulnerability model: how does stress impact on mental illness at the level of the brain and what are the consequences?', *Psychiatria Danubina*, 2010, 22(2), pp. 198–202

46. Gelkopf, M., S. Kreitler and M. Sigal, 'Laughter in a psychiatric ward: somatic, emotional, social, and clinical influences on schizophrenic patients', *Journal of Nervous and Mental Disease*, 1993, 181(5), pp. 283–9

## Chapter 7: The Dark Side of Happiness

1. Flett, G. L., K. R. Blankstein and T. R. Martin, 'Procrastination, negative self-evaluation, and stress in depression and anxiety', in J. R. Ferrari, J. H. Johnson and W. G. McCown (eds), *Procrastination and Task Avoidance*, Springer, 1995, pp. 137–67

2. Sørensen, L. B., et al., 'Effect of sensory perception of foods on appetite and food intake: a review of studies on humans', *International Journal of Obesity*, 2003, 27(10), p. 1152

3. Myers Ernst, M. and L. H. Epstein, 'Habituation of responding for food in humans', *Appetite*, 2002, 38(3), pp. 224–34

4. Brennan, P., H. Kaba and E. B. Keverne, 'Olfactory recognition: a simple memory system', *Science*, 1990, 250(4985), pp. 1223–6

5. Maldarelli, C., 'Here's why twin studies are so important to science and NASA', *Popular Science*, 1 March 2016, popsci.com

6. Kendler, K. S., et al., 'A Swedish national twin study of lifetime major depression', *American Journal of Psychiatry*, 2006, 163(1), pp. 109–14

7. Kensinger, E. A. and S. Corkin, 'Two routes to emotional memory: distinct neural processes for valence and arousal', *Proceedings of the National Academy of Sciences of the United States of America*, 2004, 101(9), pp. 3310–15

8. Hoffmann, H., E. Janssen and S. L. Turner, 'Classical conditioning of sexual arousal in women and men: effects of varying awareness and biological relevance of the conditioned stimulus', *Archives of Sexual Behavior*, 2004, 33(1), pp. 43–53

9. Dusenbury, L., et al., 'A review of research on fidelity of implementation: implications for drug abuse prevention in school settings', *Health Education Research*, 2003, 18(2), pp. 237–56

10. Freeman, B., S. Chapman and M. Rimmer, 'The case for the

plain packaging of tobacco products', *Addiction*, 2008, 103(4), pp. 580–90

11. Christiano, A. and A. Neimand, 'Stop raising awareness already', *Stanford Social Innovation Review*, Spring 2017

12. Marteau, T. M., G. J. Hollands and P. C. Fletcher, 'Changing human behavior to prevent disease: the importance of targeting automatic processes', *Science*, 2012, 337(6101), p. 1492

13. Dolcos, F., K. S. LaBar and R. Cabeza, 'Dissociable effects of arousal and valence on prefrontal activity indexing emotional evaluation and subsequent memory: an event-related fMRI study', *NeuroImage*, 2004, 23(1), pp. 64–74

14. Volkow, N. D., G.-J. Wang and R. D. Baler, 'Reward, dopamine and the control of food intake: implications for obesity', *Trends in Cognitive Sciences*, 2011, 15(1), pp. 37–46

15. Petty, R. E. and P. Brinol, 'Attitude change', *Advanced Social Psychology*, 2010, pp. 217–59

16. Beck, J. G. and S. F. Coffey, 'Assessment and treatment of PTSD after a motor vehicle collision: empirical findings and clinical observations,' *Professional Psychology: Research and Practice*, 2007, 38(6), pp. 629–39

17. Clark, R. E. and L. R. Squire, 'Classical conditioning and brain systems: the role of awareness', *Science*, 1998, 280(5360), pp. 77–81

18. Sharot, T., *The Optimism Bias: A Tour of the Irrationally Positive Brain*, Vintage, 2011

19. Cummins, R. A. and H. Nistico, 'Maintaining life satisfaction: the role of positive cognitive bias', *Journal of Happiness Studies*, 2002, 3(1), pp. 37–69

20. Sharot, T., et al., 'Neural mechanisms mediating optimism bias', *Nature*, 2007, 450(7166), pp. 102–5

21. Koob, G. F. and M. Le Moal, 'Plasticity of reward neurocircuitry and the "dark side" of drug addiction', *Nature Neuroscience*, 2005, 8(11), pp. 1442–4

22. Arias-Carrion, O. and E. Poppel, 'Dopamine, learning, and reward-seeking behavior', *Acta Neurobiologiae Experimentalis*, 2007, 67(4), pp. 481–8

23. Koob, G. F. and M. Le Moal, 'Addiction and the brain antireward system', *Annual Review of Psychology*, 2008, 59, pp. 29–53

24. Gardner, E. L., 'Introduction: addiction and brain reward and anti-reward pathways', *Advances in Psychosomatic Medicine*, 2011, 30, pp. 22–60

25. Arató, M., et al., 'Elevated CSF CRF in suicide victims', *Biological Psychiatry*, 25(3), pp. 355–9

26. Knoll, A. T. and W. A. Carlezon, 'Dynorphin, stress, and depression',

*Brain Research*, 2010, 1314C, p. 56

27. Koob, G. F. and M. L. Moal, 'Drug abuse: hedonic homeostatic dysregulation', *Science*, 1997, 278(5335), p. 52

28. 'A tale of anxiety and reward – the role of stress and pleasure in addiction relapse', *The Brain Bank North West*, 2014, thebrainbank. scienceblog.com

29. Michl, P., et al., 'Neurobiological underpinnings of shame and guilt: a pilot fMRI study', *Social Cognitive and Affective Neuroscience*, 2014, 9(2), pp. 150–7

30. Chang, Luke J., et al., 'Triangulating the neural, psychological, and economic bases of guilt aversion', *Neuron*, 2011, 70(3), pp. 560–72

31. Gilovich, T., V. H. Medvec and K. Savitsky, 'The spotlight effect in social judgment: an egocentric bias in estimates of the salience of one's own actions and appearance', *Journal of Personality and Social Psychology*, 2000, 78(2), p. 211

32. Silani, G., et al., 'Right supramarginal gyrus is crucial to overcome emotional egocentricity bias in social judgments', *Journal of Neuroscience*, 2013, 33(39), pp. 15466–76

33. Wolpert, S., 'Brain reacts to fairness as it does to money and chocolate, study shows', *UCLA Newsroom*, 21 April 2008

34. Tabibnia, G. and M. D. Lieberman, 'Fairness and cooperation are rewarding', *Annals of the New York Academy of Sciences*, 2007, 1118(1), pp. 90–101

35. Denke, C., et al., 'Belief in a just world is associated with activity in insula and somatosensory cortices as a response to the perception of norm violations', *Social Neuroscience*, 2014, 9(5), pp. 514–21

36. Blackwood, N., et al., 'Self-responsibility and the self-serving bias: an fMRI investigation of causal attributions', *NeuroImage*, 2003, 20(2), pp. 1076–85

37. O'Connor, Z., 'Colour psychology and colour therapy: caveat emptor', *Color Research and Application*, 2011, 36(3), pp. 229–34

38. Utevsky, A. V. and M. L. Platt, 'Status and the brain', *PLOS Biology*, 2014, 12(9), p. e1001941

39. Costandi, M., 'The brain boasts its own social network', *Scientific American*, 20 April 2017

40. Gil, M., et al., 'Social reward: interactions with social status, social communication, aggression, and associated neural activation in the ventral tegmental area', *European Journal of Neuroscience*, 2013, 38(2), pp. 2308–18

41. Samson, A. C. and J. J. Gross, 'Humour as emotion regulation: the differential consequences of negative versus positive humour', *Cognition and Emotion*, 2012, 26(2), pp. 375–84

42. Isenberg, D. J., 'Group polarization: a critical review and meta-analysis', *Journal of Personality and Social Psychology*, 1986, 50(6), p. 1141

43. Scheepers, D., et al., 'The neural correlates of in-group and self-face perception: is there overlap for high identifiers? ', *Frontiers in Human Neuroscience*, 2013, 7, p. 528

44. Murphy, J. M., et al., 'Depression and anxiety in relation to social status: a prospective epidemiologic study', *Archives of General Psychiatry*, 1991, 48(3), pp. 223–9

45. De Dreu, C. K., et al., 'Oxytocin promotes human ethnocentrism', *Proceedings of the National Academy of Sciences*, 2011, 108(4), pp. 1262–6

46. Hart, A. J., et al., 'Differential response in the human amygdala to racial outgroup vs ingroup face stimuli', *NeuroReport*, 2000, 11(11), pp. 2351–4

47. Avenanti, A., A. Sirigu and S. M. Aglioti, 'Racial bias reduces empathic sensorimotor resonance with other-race pain', *Current Biology*, 2010, 20(11), pp. 1018–22

48. Zebrowitz, L. A., B. White and K. Wieneke, 'Mere exposure and racial prejudice: exposure to other-race faces increases liking for strangers of that race', *Social Cognition*, 2008, 26(3), pp. 259–75

49. Rupp, H. A. and K. Wallen, 'Sex differences in response to visual sexual stimuli: a review', *Archives of Sexual Behavior*, 2008, 37(2), pp. 206–18

50. Cummins, R. G., 'Excitation transfer theory', *International Encyclopedia of Media Effects*, 2017, pp. 1–9

51. Blaszczynski, A. and L. Nower, 'A pathways model of problem and pathological gambling', *Addiction*, 2002, 97(5), pp. 487–99

52. De Brabander, B., et al., 'Locus of control, sensation seeking, and stress', *Psychological Reports*, 1996, 79(3 Pt 2), pp. 1307–12

53. Patoine, B., 'Desperately seeking sensation: fear, reward, and the human need for novelty', *The Dana Foundation*, 13 October 2009

54. Bouter, L. M., et al., 'Sensation seeking and injury risk in downhill skiing', *Personality and Individual Differences*, 1988, 9(3), pp. 667–73

55. McCutcheon, K., 'Haemophobia', *Journal of Perioperative Practice*, 2015, 25(3), p. 31

56. Burnett, D., 'James Foley's murder, and the psychology of our fascination with the gruesome', *Telegraph*, 20 August 2014

57. Varma-White, K., 'Morbid curiosity: why we can't look away from tragic images', *TODAY.com*, 19 July 2014

58. Brakoulias, V., et al., 'The characteristics of unacceptable/taboo thoughts in obsessive-compulsive disorder', *Comprehensive Psychiatry*, 2013, 54(7), pp. 750–7

59. Roberts, P., 'Forbidden thinking', *Psychology Today*, 1 May 1995
60. Johnson-Laird, P. N., 'Mental models and human reasoning', *Proceedings of the National Academy of Sciences*, 2010, 107(43), pp. 18243–50
61. Wegner, D. M., et al., 'Paradoxical effects of thought suppression', *Journal of Personality and Social Psychology*, 1987, 53(1), pp. 5–13
62. Mann, T. and A. Ward, 'Forbidden fruit: does thinking about a prohibited food lead to its consumption?', *International Journal of Eating Disorders*, 2001, 29(3), pp. 319–27
63. Etchells, P. J., et al., 'Prospective investigation of video game use in children and subsequent conduct disorder and depression using data from the Avon longitudinal study of parents and children', *PLOS ONE*, 2016, 11(1), p. e0147732

## Chapter 8: Happiness Through the Ages

1. Burnett, D., 'Women and yogurt: what's the connection? ', *Guardian*, 30 August 2013
2. Straus, W. Jr and A. J. E. Cave, 'Pathology and the posture of Neanderthal man', *Quarterly Review of Biology*, 1957, 32(4), pp. 348–63
3. Lee, M., 'Why are babies' heads so large in proportion to their body sizes? ', livestrong.com, 13 June 2017
4. Barras, C., 'The real reasons why childbirth is so painful and dangerous', bbc.com, 22 December 2016
5. Shonkoff, J. P. and D. A. Phillips (eds), 'From neurons to neighborhoods: the science of early childhood development', National Research Council and Institute of Medicine, 2000
6. Harlow, H. F., 'Love in infant monkeys', *Scientific American*, 1959
7. Houston, S. M., M. M. Herting and E. R. Sowell, 'The neurobiology of childhood structural brain development: conception through adulthood', *Current Topics in Behavioral Neurosciences*, 2014, 16, pp. 3–17
8. Stafford, T., 'Why all babies love peekaboo', bbc.com, 18 April 2014
9. Center on the Developing Child, 'Five numbers to remember about early childhood development', 2009, www.developingchild.harvard.edu
10. Dahl, R. E., 'Sleep and the developing brain', *Sleep*, 2007, 30(9), pp. 1079–80
11. Danese, A. and B. S. McEwen, 'Adverse childhood experiences, allostasis, allostatic load, and age-related disease', *Physiology and Behavior*, 2012, 106(1), pp. 29–39
12. Shonkoff, J. P., et al., 'The lifelong effects of early childhood

adversity and toxic stress', *Pediatrics*, 2012, 129(1), pp. e232–46

13. Avants, B., et al. 'Early childhood home environment predicts frontal and temporal cortical thickness in the young adult brain', Society for Neuroscience annual meeting, 2012

14. Jack, F., et al., 'Maternal reminiscing style during early childhood predicts the age of adolescents' earliest memories', *Child Development*, 2009, 80(2), pp. 496–505

15. Brink, T. T., et al., 'The role of orbitofrontal cortex in processing empathy stories in four- to eight-year-old children', *Frontiers in Psychology*, 2011, 2, p. 80

16. Neisser, U., et al., 'Intelligence: knowns and unknowns', *American Psychologist*, 1996, 51(2), p. 77

17. Sherif, M., et al., *Intergroup Conflict and Cooperation: The Robbers Cave Experiment*, Wesleyan, 1954/1961

18. Houston, S. M., et al., 'The neurobiology of childhood structural brain development: conception through adulthood', *Current Topics in Behavioral Neurosciences*, 2014, 16, pp. 3–17

19. Galbally, M., et al., 'The role of oxytocin in mother–infant relations: a systematic review of human studies', *Harvard Review of Psychiatry*, 2011, 19(1), pp. 1–14

20. Wan, M. W., et al., 'The neural basis of maternal bonding', *PLOS ONE*, 2014, 9(3), p. e88436

21. Magon, N. and S. Kalra, 'The orgasmic history of oxytocin: love, lust, and labor', *Indian Journal of Endocrinology and Metabolism*, 2011, 15(7), p. 156

22. Noriuchi, M., Y. Kikuchi and A. Senoo, 'The functional neuroanatomy of maternal love: mother's response to infant's attachment behaviors', *Biological Psychiatry*, 2008, 63(4), pp. 415–23

23. Schore, A. N., 'Effects of a secure attachment relationship on right brain development, affect regulation, and infant mental health', *Infant Mental Health Journal*, 2001, 22(1–2), pp. 7–66

24. Ainsworth, M. D. S., et al., *Patterns of Attachment: A Psychological Study of the Strange Situation*, Psychology Press, 2015

25. Wiseman, H., O. Mayseless and R. Sharabany, 'Why are they lonely? Perceived quality of early relationships with parents, attachment, personality predispositions and loneliness in first-year university students', *Personality and Individual Differences*, 2006, 40(2), pp. 237–48

26. Blustein, D. L., M. S. Prezioso and D. P. Schultheiss, 'Attachment theory and career development', *The Counseling Psychologist*, 1995, 23(3), pp. 416–32

27. Potard, C., et al., 'The relationship between parental attachment and sexuality in early adolescence', *International Journal of Adolescence*

*and Youth*, 2017, 22(1), pp. 47–56

28. Baumrind, D., 'The influence of parenting style on adolescent competence and substance use', *Journal of Early Adolescence*, 1991, 11(1), pp. 56–95

29. Haycraft, E. and J. Blissett, 'Eating disorder symptoms and parenting styles', *Appetite*, 2010, 54(1), pp. 221–224

30. Baumrind, D., 'Current patterns of parental authority', *Developmental Psychology*, 1971, 4(1 pt 2), p. 1

31. Foster, A. D. and M. R. Rosenzweig, 'Learning by doing and learning from others: human capital and technical change in agriculture', *Journal of Political Economy*, 1995, 103(6), pp. 1176–1209

32. Landry, S. H., et al., 'Does early responsive parenting have a special importance for children's development or is consistency across early childhood necessary?', *Developmental Psychology*, 2001, 37(3), pp. 387–403

33. Kaplowitz, P. B., et al., 'Earlier onset of puberty in girls: relation to increased body mass index and race', *Pediatrics*, 2001, 108(2), p. 347

34. Neubauer, A. C. and A. Fink, 'Intelligence and neural efficiency: measures of brain activation versus measures of functional connectivity in the brain', *Intelligence*, 2009, 37(2), pp. 223–9

35. Santos, E. and C. A. Noggle, 'Synaptic pruning', in S. Goldstein and J. A. Naglieri (eds), *Encyclopedia of Child Behavior and Development*, Springer, 2011, pp. 1464–5

36. Carskadon, M. A., 'Patterns of sleep and sleepiness in adolescents', *Pediatrician*, 1990, 17(1), pp. 5–12

37. Owens, J. A., K. Belon and P. Moss, 'Impact of delaying school start time on adolescent sleep, mood, and behavior', *Archives of Pediatrics and Adolescent Medicine*, 2010, 164(7), pp. 608–14

38. McClintock, M. K. and G. Herdt, 'Rethinking puberty: the development of sexual attraction', *Current Directions in Psychological Science*, 1996, 5(6), pp. 178–83

39. Casey, B. J., R. M. Jones and T. A. Hare, 'The adolescent brain', *Annals of the New York Academy of Sciences*, 2008, 1124(1), pp. 111–26

40. Spear, L. P., 'The adolescent brain and age-related behavioral manifestations', *Neuroscience and Biobehavioral Reviews*, 2000, 24(4), pp. 417–63

41. Reyna, V. F. and F. Farley, 'Risk and rationality in adolescent decision making: implications for theory, practice, and public policy', *Psychological Science in the Public Interest*, 2006, 7(1), pp. 1–44

42. Lenroot, R. K. and J. N. Giedd, 'Brain development in children and adolescents: insights from anatomical magnetic resonance imaging', *Neuroscience and Biobehavioral Reviews*, 2006, 30(6), pp. 718–29

43. Henry, J. P., 'Biological basis of the stress response', *Integrative Physiological and Behavioral Science*, 1992, 27(1), pp. 66–83

44. Philpot, R. M. and L. Wecker, 'Dependence of adolescent novelty-seeking behavior on response phenotype and effects of apparatus scaling', *Behavioral Neuroscience*, 2008, 122(4), pp. 861–75

45. Walter, C., *Last Ape Standing: The Seven-Million-Year Story of How and Why We Survived*, Bloomsbury Publishing USA, 2013

46. Weon, B. M. and J. H. Je, 'Theoretical estimation of maximum human lifespan', *Biogerontology*, 2009, 10(1), pp. 65–71

47. Deng, W., J. B. Aimone and F. H. Gage, 'New neurons and new memories: how does adult hippocampal neurogenesis affect learning and memory?', *Nature Reviews Neuroscience*, 2010, 11(5), pp. 339–50

48. Rakic, P., 'Neurogenesis in adult primate neocortex: an evaluation of the evidence', *Nature Reviews Neuroscience*, 2002, 3(1), pp. 65–71

49. Shephard, E., G. M. Jackson and M. J. Groom, 'Learning and altering behaviours by reinforcement: neurocognitive differences between children and adults', *Developmental Cognitive Neuroscience*, 2014, 7: pp. 94–105

50. Nisbett, R. E., et al., 'Intelligence: new findings and theoretical developments', *American Psychologist*, 2012, 67(2), pp. 130–59

51. Esch, T. and G. B. Stefano, 'The neurobiology of stress management', *Neuroendocrinology Letters*, 2010, 31(1), pp. 19–39

52. Goh, C. and M. Agius, 'The stress-vulnerability model: how does stress impact on mental illness at the level of the brain and what are the consequences?', *Psychiatria Danubina*, 2010, 22(2), pp. 198–202

53. Ulrich-Lai, Y. M., et al., 'Pleasurable behaviors reduce stress via brain reward pathways', *Proceedings of the National Academy of Sciences of the United States of America*, 2010, 107(47), pp. 20529–34

54. Milman, A., 'The impact of tourism and travel experience on senior travelers' psychological well-being', *Journal of Travel Research*, 1998, 37(2), pp. 166–70

55. Glocker, M. L., et al., 'Baby schema in infant faces induces cuteness perception and motivation for caretaking in adults', *Ethology*, 2009, 115(3), pp. 257–63

56. 'Holly Brockwell', from www.hollybrockwell.com

57. Brockwell, H., 'Why can't I get sterilised in my 20s?', *Guardian*, 28 January 2015

58. Feldman, S., 'Structure and consistency in public opinion: the role of core beliefs and values', *American Journal of Political Science*, 1988, pp. 416–40

59. Moussavi, S., et al., 'Depression, chronic diseases, and decrements in health: results from the World Health Surveys', *Lancet*, 2007, 370(9590), pp. 851–8

60. Pinquart, M., 'Creating and maintaining purpose in life in old age: a meta-analysis', *Ageing International*, 2002, 27(2), pp. 90–114

61. Bonanno, G. A., et al., 'Resilience to loss and chronic grief: a prospective study from preloss to 18-months postloss', *Journal of Personality and Social Psychology*, 2002, 83(5), p. 1150

62. Chang, S. H. and M. S. Yang, 'The relationships between the elderly loneliness and its factors of personal attributes, perceived health status and social support', *Kaohsiung Journal of Medical Sciences*, 1999, 15(6), pp. 337–47

63. Peters, R., 'Ageing and the brain', *Postgraduate Medical Journal*, 2006, 82(964), pp. 84–8

64. Myers, B. L. and P. Badia, 'Changes in circadian rhythms and sleep quality with aging: mechanisms and interventions', *Neuroscience and Biobehavioral Reviews*, 1996, 19(4), pp. 553–71

65. Whalley, L. J., 'Brain ageing and dementia: what makes the difference?', *British Journal of Psychiatry*, 2002, 181(5), p. 369

66. Ebner, N. C. and H. Fischer, 'Emotion and aging: evidence from brain and behavior', *Frontiers in Psychology*, 2014, 5, p. 996

67. Chapman, S. B., et al., 'Shorter term aerobic exercise improves brain, cognition, and cardiovascular fitness in aging', *Frontiers in aging neuroscience*, 2013, 5

68. Almeida, R. P., et al., 'Effect of cognitive reserve on age-related changes in cerebrospinal fluid biomarkers of Alzheimer disease', *JAMA Neurology*, 2015, 72(6), pp. 699–706

69. 'Elderly playgrounds', *Injury Prevention*, 2006, 12(3), p. 170

70. Sharot, T., *The Optimism Bias: A Tour of the Irrationally Positive Brain*, Vintage, 2011

71. Burnett, D., '"Your film has ruined my childhood!" Why nostalgia trumps logic on remakes', *Guardian*, 1 June 2016

72. Sedikides, C. and T. Wildschut, 'Past forward: nostalgia as a motivational force', *Trends in Cognitive Sciences*, 2016, 20(5), pp. 319–21

73. Zhou, X., et al., 'Counteracting loneliness', *Psychological Science*, 2008, 19(10), pp. 1023–9

74. Caspari, R., 'The evolution of grandparents', *Scientific American*, 2011, 305(2), pp. 44–9

75. Jago, C., 'Always Look on the Bright Side of Death', 2017, http://rationalcancer.blogspot.com/

# Index